Feminist Spaces

Feminist Spaces introduces students and ac and empirical studies in feminist geograph, research including: embodiment, sexuality, masculinity, intersectional analysis, and environment and development. In addition to considering gender as a primary subject, this book provides a comprehensive overview of feminist geography by highlighting contemporary research conducted from a feminist framework which goes beyond the theme of gender to include issues such as social justice, activism, (dis)ability, and critical pedagogy.

Through case studies, this book challenges the construction of dichotomies that tend to oversimplify categories such as developed and developing, urban and rural, and the Global North and South, without accounting for the fluid and intersecting aspects of gender, space, and place. The chapters weave theoretical and empirical material together to meet the needs of students new to feminism, as well as those with a feminist background but new to geography, through attention to basic geographical concepts in the opening chapter. The text encourages readers to think of feminist geography as addressing not only gender, but a set of methodological and theoretical perspectives applied to a range of topics and issues. A number of interactive exercises, activities, and 'boxes' or case studies, illustrate concepts and supplement the text. These prompts encourage students to explore and analyze their own positionality, as well as motivate them to change and impact their surroundings.

Feminist Spaces emphasizes activism and critical engagement with diverse communities to recognize this tradition within the field of feminism, as well as within the discipline of geography. Combining theory and practice as a central theme, this text will serve graduate level students as an introduction to the field of feminist geography, and will be of interest to students in related fields such as environmental studies, development, and women's and gender studies.

Ann M. Oberhauser is Professor of Sociology and Director of Women's and Gender Studies, Iowa State University, USA.

Jennifer L. Fluri is Associate Professor, Department of Geography, University of Colorado-Boulder, USA.

Risa Whitson is Associate Professor, Department of Geography and Women's, Gender, and Sexuality Studies Program, Ohio University, USA.

Sharlene Mollett is Assistant Professor, Department of Human Geography and Centre for Critical Development Studies, University of Toronto, Canada.

Feminist Spaces

Gender and Geography in a Global Context

Ann M. Oberhauser, Jennifer L. Fluri, Risa Whitson, and Sharlene Mollett

LONDON AND NEW YORK

First published 2018
by Routledge
2 Park Square, Milton Park, Abingdon, Oxon OX14 4RN

and by Routledge
711 Third Avenue, New York, NY 10017

Routledge is an imprint of the Taylor & Francis Group, an informa business

© 2018 Ann M. Oberhauser, Jennifer L. Fluri, Risa Whitson, and Sharlene Mollett

The right of Ann M. Oberhauser, Jennifer L. Fluri, Risa Whitson, and Sharlene Mollett to be identified as authors of this work has been asserted by them in accordance with sections 77 and 78 of the Copyright, Designs and Patents Act 1988.

All rights reserved. No part of this book may be reprinted or reproduced or utilised in any form or by any electronic, mechanical, or other means, now known or hereafter invented, including photocopying and recording, or in any information storage or retrieval system, without permission in writing from the publishers.

Trademark notice: Product or corporate names may be trademarks or registered trademarks, and are used only for identification and explanation without intent to infringe.

British Library Cataloguing-in-Publication Data
A catalogue record for this book is available from the British Library

Library of Congress Cataloging-in-Publication Data
A catalog record for this book has been requested

ISBN: 978-1-138-92452-9 (hbk)
ISBN: 978-1-138-92453-6 (pbk)
ISBN: 978-1-315-68427-7 (ebk)

Typeset in Bembo
by Keystroke, Neville Lodge, Tettenhall, Wolverhampton

Contents

List of figures	vii
List of boxes	ix
About the authors	x
Acknowledgments	xii

1 ENGAGING FEMINIST SPACES: INTRODUCTION AND OVERVIEW   1
Ann M. Oberhauser, Jennifer L. Fluri, Risa Whitson, and Sharlene Mollett

2 THE BODY, PERFORMANCE, AND SPACE   25
Jennifer L. Fluri

3 SPACES OF CULTURE AND IDENTITY PRODUCTION: HOME, CONSUMPTION, AND THE MEDIA   47
Risa Whitson

4 GENDERING THE RIGHT TO THE CITY   77
Risa Whitson

5 GENDERED WORK AND ECONOMIC LIVELIHOODS   107
Ann M. Oberhauser

6 FEMINIST POLITICAL GEOGRAPHY AND GEOPOLITICS   131
Jennifer L. Fluri

7 ENVIRONMENTAL STRUGGLES ARE FEMINIST STRUGGLES: FEMINIST POLITICAL ECOLOGY AS DEVELOPMENT CRITIQUE   155

Sharlene Mollett

8 FEMINIST SPACES: CONCLUSION AND REFLECTIONS   189

Sharlene Mollett, Jennifer L. Fluri, Risa Whitson, and Ann M. Oberhauser

Bibliography   196
Index   223

peace during political transition (2014–2016) and was funded by the US Institute of Peace.

Risa Whitson is Associate Professor, Department of Geography and Women's, Gender, and Sexuality Studies Program, Ohio University. Her research focuses on the social geographies of gender, work, and leisure. She is particularly interested in how non-standard labor relations, and in particular informal work, constitute an important element of changing economic structures and gender performances. The majority of her research has taken place in the context of contemporary Argentina, although she has also conducted research in a US context. She has published numerous articles in journals that include *Social & Cultural Geography*, *Antipode*, and *Annals of the Association of American Geographers*. As a jointly appointed faculty member in Geography and Women's, Gender, and Sexuality Studies, she has taught upper-level courses in social geography, gender and development, and global feminisms for over a decade.

Sharlene Mollett is Assistant Professor, Department of Human Geography and Centre for Critical Development Studies, University of Toronto. Her research interrogates the ways in which gender, race, and property rights are co-constituted and imbued in international development policies and practices in Central America. Drawing upon the insights from feminist and post-colonial political ecologies, and with a sustained look at protected area management and residential tourism in Honduras and Panama respectively, her work unpacks the ways in which indigenous, Afro-descendant, and Mestizo communities make claims to, and are displaced from rights to, land and territorial control. Her published work has featured in such journals as *Gender, Place & Culture*, *Annals of the Association of American Geographers*, *Antipode*, *Latin American Research Review*, and *Geoforum*. She brings her research insights to her undergraduate and graduate teaching in geography and development studies.

Acknowledgments

This book builds on the legacy of scholarship and praxis that began in the field of feminist geography several decades ago. We are grateful for and acknowledge the landmark contributions by scholars and activists whose efforts are reflected in this foundational body of work. The discussions in this book seek to offer not only an entrée into, but also fresh insight on, the rich theoretical and empirical work present in the discipline today. This work challenges us to think beyond popular and mainstream ways of talking about diversity, inclusion, praxis, and marginalization. While this book should not be read as an exhaustive survey of feminist geographic scholarship, we hope that the perspectives and dialogues in this book will lead to an expanded understanding of the contested and contextual futures that we, as feminists, face in the 21st century.

Many people and communities contributed to this project, and we especially benefitted from the tremendous support and feedback of colleagues, students, friends, and family. Reviewers of the manuscript were instrumental at various stages of the writing process. Anonymous reviewers provided crucial insights to the initial proposal, as well as to early versions of the manuscript as a whole. We especially appreciate the editorial support of Katie Gillespie in preparing the first copy of the manuscript and Deborah Burns who was extremely helpful in organizing the details and preparing material for the final submission of the book.

This project was also facilitated by the generous hospitality of several people, including Martha Scott who shared her apartment in Chicago with us while we worked on the manuscript. Funding for this project was provided by the College of Liberal Arts and Sciences at Iowa State University, the Center for International Studies at Ohio University, and the University of Colorado-Boulder.

Figures

1.1	Slutwalk Toronto, 2012	2
1.2	The Castro District in San Francisco, California	8
1.3	Free speech zone at Iowa State University	10
1.4	Trans March during the Gay Pride Parade in San Francisco, California	20
2.1	'We Can Do It!' poster for Westinghouse	29
3.1	Disneyland's Main Street, USA	51
3.2	Shopping malls as feminized spaces	59
3.3	An open-air market in Rosario, Santa Fe, Argentina	62
3.4	Galerías Pacífico mall in downtown Buenos Aires	62
3.5	Bars are examples of public leisure spaces where gendered and sexualized meaning is produced	63
3.6	Fans at a Philadelphia Eagles game	65
3.7	The logo of the Women2Drive movement	71
3.8	The logo of Feminist Frequency	71
4.1	A march to protest police brutality during the second week of Occupy Wall Street in New York City	81
4.2	Black Lives Matter activists protesting in Minneapolis, Minnesota	81
4.3	Central Park in New York City	83
4.4	Hashemite Plaza in Amman, Jordan	86
4.5	A mental map of spaces of fear produced by female college students	89
4.6	A mental map of spaces of fear produced by female college students	89

Figures

4.7	Walking-distance access to infrastructure for social reproduction in Madrid	94
4.8	Global measure of access to infrastructure for social reproduction in Madrid	95
4.9	Jardin du Luxembourg in Paris, France	102
5.1	A woman selling farm produce in Limpopo Province, South Africa, 2012	111
5.2	Women as a percentage of total employed in selected occupations, 2011	116
5.3	Occupational stereotypes: the case of nursing	117
5.4	Workers at the Mustang socks factory in Palghar, India	122
6.1	Ethnolinguistic groups in the Caucasus region	135
6.2	An ice cream seller at Qarghar Lake, Afghanistan	149
7.1	Berta Cáceres, a Lenca and Women Human Rights Defender from Honduras	156
7.2	The Chipko Movement	159
7.3	Women farmers in Rajkot, Gujarat, India	160
7.4	Investing in girls: is it this easy?	167
7.5	Rio Plátano Biosphere Reserve map	169
7.6	A Miskito woman farmer	170
7.7	Land grabbing in Myanmar	173
7.8	In defense of human rights	175
7.9	Yemenja celebrations	177
7.10	Water procurement in New Delhi, India	179
7.11	US veterans join Standing Rock protest	182
7.12	Resilience in the face of climate change in Kenya	185

Boxes

1.1	Free speech, safe space, and campus climate	10
1.2	Contested views of development	14
2.1	Examples of changing gender norms	29
2.2	Transgender transformations attempting to queer gender boundaries	34
2.3	Miss Landmine and Aimee Mullins	38
2.4	The corporeal marker assignment	44
3.1	The politics of culture	51
3.2	Gendered images of the home	54
3.3	Gendered spaces in the home	55
3.4	Changing gender norms through new media activism	71
4.1	Who has the right to your city?	81
4.2	Fear in public space	89
4.3	Access to infrastructure for social reproduction in Madrid	94
4.4	The Right to Pee campaign	96
5.1	Women in the informal sector	111
5.2	Women's economic empowerment	120
6.1	The lives and deaths of the civil rights movement in the US	141
6.2	Female Engagement Teams (FETs)	148
6.3	Everyday militarization	149

About the authors

Ann M. Oberhauser is Professor of Sociology and Director of Women's and Gender Studies, Iowa State University. She has conducted research on gender and economic livelihoods with an emphasis on the informal sector and collective economic strategies. Much of her work has taken place in Appalachia and sub-Saharan Africa, where she examines economic strategies among both rural and urban households and communities. She is the author of numerous articles and book chapters published in journals such as *Gender, Place & Culture, African Geographical Review, Antipode*, and *GeoJournal*. She recently co-edited a book with Ibipo Johnston-Anumonwo titled *Global Perspectives on Gender and Space: Engaging Feminism and Development*. She has taught courses on gender, development, and feminist geography to both graduate and undergraduate students at several US and international universities.

Jennifer L. Fluri is Associate Professor, Department of Geography, University of Colorado-Boulder. Her research is regionally focused in South Asia with an emphasis on Afghanistan. Her post-doctoral research addressed the geopolitics, gender politics, and geo-economics of international military, aid, and development interventions in Afghanistan. Her research has appeared in various journals including *Annals of the Association of American Geographers, Transactions of the Institute of British Geographers*, and *Political Geography*. She recently completed a co-authored manuscript about the intimate geopolitics of aid and development titled *Carpetbaggers of Kabul and Other American–Afghan Entanglements: Intimate Development and the Currency of Gender and Grief*, published in 2017 by the University of Georgia Press. Her latest project examines Afghan women's organizations. This project analyzes women's efforts toward

Acknowledgments

Ann Oberhauser would like to acknowledge the support and insights of several colleagues at West Virginia University where the seeds of this project were planted. In particular, she would like to extend her gratitude to Karen Culcasi, Cynthia Gorman, Maria Perez, and Bradley Wilson for their feedback on early drafts of the book. At Iowa State University, she has received invaluable support from students and colleagues in the Women's and Gender Studies Program. She would also like to thank her mother, who raised a family and, after reading Betty Friedan, was inspired to continue her career in counseling. She is also indebted to Benzel and Frances who taught her the importance of mindful living.

Jennifer Fluri would like to thank her research participants and collaborators, particularly Rachel Lehr and Najibullah Sedeqe. Many thanks are extended to her co-authors for their insights, assistance, comments, and suggestions. As always, thanks to her husband Jeff and daughters Jessica and Samantha for their continual support and love.

Risa Whitson would like to thank Susan Ngbabare and Sergina Loncle for their work on developing and organizing bibliographic material during the early preparation of the book. Thanks also to Anushka Gole for her assistance with bibliographic work and her feedback and ideas on the chapter content and illustrations. She would like to express gratitude to her co-authors who provided critical and thoughtful feedback, support, and inspiration during the writing process. A special thanks to Woody Wood for the beautiful photographs he took to accompany this text, as well as for his encouragement and advice during the writing process.

Sharlene Mollett would like to thank all her co-authors for their insights and support during the writing process. She would also like to thank the Miskito communities of Honduras for sharing their stories of struggle and celebration. She sends 'big' thanks to her undergraduate students at the University of Toronto, Scarborough, for the ways in which so many of them so enthusiastically read and re-tell the ethnographic examples from her research and the work of others. She has written her chapter with these students in mind. And of course, thanks to Zavier for being a source of sunshine every day—lots of love.

Finally, we are grateful to the community of feminist geographers and other critical scholars who have nurtured and pushed us to continue this engaging, but also challenging, work of writing a co-authored book. The Geographic Perspectives on Women (GPOW), the International Geographical Union's Commission on Gender and Geography, the Feminist Geography Group, and many other communities gave us invaluable support during the writing of this book.

Chapter 1

ENGAGING FEMINIST SPACES
INTRODUCTION AND OVERVIEW

Ann M. Oberhauser, Jennifer L. Fluri,
Risa Whitson, and Sharlene Mollett

Introduction

On January 24, 2011, during a safety forum at York University's Osgoode Hall Law School located in Toronto, Canada, Toronto Police Constable, Michael Sanguinetti, advised students on how to prevent sexual assault. The constable instructed that 'women should avoid dressing like sluts' (Millar 2011). This comment and the mounting public reactions sparked a protest called SlutWalk that quickly moved from Toronto throughout North America and around the globe. SlutWalk protests became infamous because protestors wore clothes that resembled the stereotypical image of 'sluts' as a strategy to challenge prevailing presumptions among law enforcement officers and the broader public: that women are raped because of how they dress (Figure 1.1). The goal of the protests was to illustrate the assertion that dressing a certain way invites rape. These protests also sought to shift the conversation away from victim blaming and toward the acts and actions of perpetrators.

Feminist geographers have participated in this transnational movement through their scholarship. For instance, Jason Lim and Alexandra Fanghanel (2013) highlight another group within one SlutWalk protest who wore hoodies and hijabs. This group sought to challenge sexist along with racist, classist, and

Figure 1.1 SlutWalk Toronto, 2012. Reproduced courtesy of Loretta Lime.

Introduction and overview

Islamophobic assumptions about dress, such as associating hoodies with criminal activity, and hijabs with religious extremism. Since veiling has become a contested form of dress in many non-Muslim majority nations in the Global North, these SlutWalk activists wore hijabs to highlight existing prejudices about veiling among Muslim women. These women further reinforced the point of the protest: that the way in which a woman dresses is not an indication of her culpability in her own assault. Such activists show solidarity against victim blaming more generally, while also countering stereotypes that justify other forms of corporeal violence based on how someone is dressed.

While SlutWalk energized some feminist circles, for others the movement garnered concern. A group of African American feminists and activists called for the need for historical reflection and offered this insight in an open letter to the movement:

> We are deeply concerned. As Black women and girls we find no space in SlutWalk, no space for participation and to unequivocally denounce rape and sexual assault as we have experienced it. We are perplexed by the use of the term 'slut' and by any implication that this word . . . should be re-appropriated . . . The way in which we are perceived and what happens to us before, during and after sexual assault crosses the boundaries of our mode of dress. Much of this is tied to our particular history. In the United States, where slavery-constructed Black female sexualities, Jim Crow kidnappings, rape and lynchings, gender misrepresentations, and more recently, where the Black female immigrant struggle combine, 'slut' has different associations for Black women . . . As Black women, we do not have the privilege or the space to call ourselves 'slut' without validating the already historically entrenched ideology and recurring messages about what and who the Black woman is.
>
> ('An Open Letter from Black Women to the SlutWalk,' September 23, 2011, from Black Women's Blueprint (BWB), cited in Brison 2011)

SlutWalk and its critics offer an instructive example to feminist geographic debates about how feminists relate to difference. The concept of difference is important because feminist or women's movements are often based on the idea that all women are the same and can bond over shared experiences such as pay equity, motherhood, and domestic violence, for example. However, our gendered experiences are not just shaped by our roles as women and men, but rather other factors of social difference play a role, such as our race, ethnicity, nationality, religion, sexuality, immigration status, age, and language, to name a few. This book seeks to contribute to a growing commitment among feminist geographers to attend to ways in which various forms of social difference shape our understanding of gender.

Intersectionality and feminist geography

This book's approach to feminist geography addresses numerous and contested power relations that shape social dynamics in space, the role of 'difference' and intersecting social identities, and how praxis and activism help to define this area of geography. Our examples and discussions underscore how feminist approaches to, and analyses of, gender and other social identities are embedded in geographic processes. Likewise, gender relations are constitutive of material and physical spaces, as well as symbolic and discursive spaces. In other words, where we go, how we get there, and our presence in certain places are influenced by, and have an impact on, social identity. Spatial dimensions of social identity such as gender, class, sexuality, and race are rooted in unequal and historically constructed power relations that privilege some people and marginalize others.

This book also emphasizes difference as a defining approach in feminist geography. Diverse social identities and histories raise important issues about how gender, class, race, ethnicity, age, sexuality, and other axes of power are formative aspects of who we are and what we believe and value. Furthermore, certain social identities grant people privileged positions as they engage in work, education, or political activities. Feminist theory and practice have examined themes surrounding difference as a means of including voices that have been traditionally excluded from mainstream feminist analyses. Some prominent examples are based on critiques of US-based feminism as representing the perspectives of white Western women only, leaving women from the Global South, as well as US-based immigrant and minority women and their priorities, outside feminist debates and programs. Therefore, this book addresses different feminisms. We use the plural 'feminisms' to highlight the diversity of feminist theories, ideas, activisms, and thoughts as they occur in various places and sociocultural, economic, and political contexts. Recognizing difference and attending to the power of differences is a theme of feminist geographic theorizing and research that we explore in this book.

We explore the theme of difference in this book through the use of the concept of **intersectionality**. Intersectionality emerged as a challenge from

Introduction and overview

US-based Black feminists who critiqued the narrow attention paid to gender and patriarchy without recognition of the messy, interwoven, and mutually constituted ways in which race, gender, sexuality, and class congeal through power and oppression. Attention to the differences of race, class, and sexuality as a way to understand gender power fundamentally decenters and disrupts a singular notion of women and feminism and requires a rethinking of how we study gender. This concept and its theorization is commonly credited to critical race theorist Kimberle Crenshaw, who has asked feminists to understand social difference as interdependent and interlocking since the 1980s. As she writes, 'I used the concept of intersectionality to denote the various ways in which race and gender interact to shape the multiple dimensions of Black women's employment experiences' (1991, 1244). While formally introduced by Crenshaw, this project of shifting our understanding of gender—away from single-axis analysis to a category shaped by multiple forms of power bound up in intersectionality—was supported by other feminist scholars. For instance, Patricia Hill Collins (2000) introduced the term 'matrix of domination' as a move to disrupt additive models of oppression grounded in the binary thinking that emanated from Eurocentric masculinist thought (225).

The collective of Black feminist thinkers at the time, including bell hooks and Audrey Lorde, as well as radical women of color Cherie Moraga and Gloria Anzaldúa among others, promoted intersectional thought as a way to make the discrimination against women of color and African American women visible, and to show how such discrimination is wedded to the ongoing prevalence of white supremacy. The emergence of this concept in the context of racial segregation (legal and *defacto*) and the institutionalized racial violence in the US and around the globe makes intersectionality deeply spatial (McKittrick 2006; Mollett and Faria forthcoming; Shabazz 2012). The growing use of the term not only contributes to feminist geographic thought and the understanding of privilege and oppression in the US, but also around the globe (see Chapter 7). Increasingly, intersectionality is used to understand broad social processes within feminist geography from social reproduction to urbanization, migration, land struggles, and environmental activism, to name a few. While a push for difference is now firmly established in feminist geography, the lines of inquiry are importantly still highlighted separately, as each builds on interwoven and separate struggles. Such entanglements leave fertile ground for feminist geographers to shape understandings of social power and avenues for social change.

The focus of this book is feminist geography as a means of contextualizing diverse social identities and activism within space and place. The term 'feminist' and the need to recognize the existence of multiple feminisms imply an inclusive and multi-layered approach that allows a critical understanding of power relations and social identities. As highlighted by Lise Nelson and Joni Seager (2005), feminism is defined by 'explicitly political commitments' against multiple oppressions and the need to assert 'the importance and salience of foregrounding women as a subject of study and "gendering" as a social and spatial process' (6–7). The discussions included in this book also emphasize the breadth of theoretical approaches in this field, which have multiplied in recent decades to include queer studies, critical race theory, and political ecology. Finally, this book builds on feminist geography's contributions to the combination of theory and practice, or praxis, and the field's overall resistance to unequal power dynamics. These contributions are found in the academic and activist realms of feminist geography as suggested in each of the chapters.

This introductory chapter invites readers to engage with themes and approaches to feminist geography in a variety of ways. We offer a look at some of feminist geography's influences and the kinds of scholarship we as a subfield take on. Attending to the diversity of feminisms, this book is written by four authors with disparate approaches, research experiences, and voices. In general, the topics presented are not exhaustive and instead speak to our particular research and teaching interests. The book also includes both brief and extended examples, and we offer practical exercises that ground the discussions aimed at undergraduates in their third and fourth years of study, as well as early graduate students in geography, women's and gender studies, feminist studies, and related fields. This approach builds on the contributions of feminist geography by encouraging readers to actively engage with the material and concepts presented here in their daily lives.

Geography/feminism/gender: concepts and context

While feminist geographic engagement with intersectionality grows strong, this has not always been the case. In this book, we trace how our intersectional understandings of gender build upon the scholarly work of women and feminist geographers who have paved the way for our contemporary insights and research in human geography.

Geography provides an important lens through which to examine the meanings of space and place, as well as the socio-spatial processes that shape our lives. The discipline of geography addresses human–environment relations and the role of economic, political, and cultural forces in shaping these relations. Many aspects of feminist geography are both informed by and embedded in these geographic concepts as a means of critically analyzing social inequality and power relations. Geography plays a key role in how gendered power relations impact our lives and the places where we work and live. These connections are evident in studies that look at workplace dynamics, urban social movements, household divisions of labor, and the use of natural resources.

Feminist perspectives on space and place

Attention to history, culture, and politics is crucial to feminist understandings of space and place. Historically, geographers have approached place as a bounded area with unique characteristics to which people attach certain identities and subjective meanings. Space has commonly been defined as an absolute, objective concept concerned with geometric properties and coordinates. Similarly, geographers generally conceive of space as an abstract dimension, much like time, through which we move and travel, and onto which patterns can be mapped. Place, however, is more frequently connected to the local and particular. Feminist geographers have challenged these conventional understandings of place and space in two primary ways. First, feminist geographers have argued that **places** are not harmonious, objective constructs with meanings and boundaries that are fixed and static. Rather, feminist geographers understand places as dynamic, socially constructed configurations with fluid and porous boundaries, which are perceived and experienced differently by different groups of people. Similarly, feminist geographers conceptualize **space** not as an abstract, natural, and neutral category, but as an open concept that is always in process, a product itself of interrelations and interaction, and as a sphere of multiplicity and heterogeneity (Massey 2005).

Doreen Massey, Gillian Rose, Susan Hanson, Linda McDowell, Audrey Kobayashi, Linda Peake, Laura Pulido, Mona Domosh, and Cindi Katz are among the pathbreaking feminist geographers who highlight how place and space are shaped and created by social identities that are, in turn, linked to different axes of power. For example, gay men experience urban neighborhoods differently than heterosexual men based on their sexuality and the predominance of normative **heterosexual spaces** of the city (Oswin 2008). In these situations, the dominant social norms found in some urban spaces exclude those who do not identify as heterosexual or perform normative heterosexual relations. In contrast, certain urban spaces are both ascribed and represented

Figure 1.2 The Castro District in San Francisco, California. The rainbow flags hung on light poles and buildings provide a visual way of marking the Castro as a gay space. Reproduced courtesy of bjaglin.

as gay spaces, such as the Castro District in San Francisco, Soho in London, and Church and Wellesley in Toronto. These neighborhoods provide a space for commercial activities, celebrations, and residential areas for the LGBTQ community (Figure 1.2).

Household and domestic spaces are a regular site of research for feminist geographers, as they question the seemingly harmonious and binary gendered relations within these spaces. Again, gender and other social identities are examined in households through a historical lens. During the early 1900s, certain tasks and spaces in the household were often 'assigned' to women and/or men such as cooking in the kitchen, repairing machinery in the garage, or working outside on the lawn. **Patriarchal norms** or conventions associated with the home labeled domestic spaces as feminized. As a result, gendered social relations contributed to the construction of the home as a 'woman's place.' As a consequence, domestic and household spaces have been interpreted by some feminists as restrictive and oppressive, precisely because they reproduce patriarchal norms and social relations. As discussed further in Chapter 3, feminist geographers' research on the home broadens conventional understanding of gender and space by examining the complexity of the home as a site of multiple experiences and expressions; from being a site of safety and security, family and comfort, to being one of immobility, oppression, and even violence.

Examining gender within geographic analyses has also included an understanding of and interrogation of place. Places are defined as areas with particular meaning for those who inhabit or use them. Corporate offices and commercial businesses, for instance, are places where people engage in work-related activities and social interactions. In many cases, power relations and hierarchies exist within these work environments among employees, employers, and managers. As discussed in Chapter 5, these relations are often reflected in and impacted by pay scales, promotional processes, and hiring practices. McDowell (2013) has conducted extensive research on the gendered construction of work and, in particular, the embodied attributes and workplace performances that conform to employer and societal expectations. Her analysis of immigrant women in the UK during the 20th and early 21st centuries demonstrates how their work experiences have been influenced by societal constructions of gender, race, and ethnicity.

Discussions of space and place within feminist geography raise related issues concerning public and private space. The divisions and separation of public and private spaces are often misunderstood. Thinking about spaces and their related activities as dichotomous or dualistic has been challenged by feminists who argue that spaces are constructed and experienced as fluid and contested. Safe spaces are generally assumed to be private homes and other living spaces, like campus dormitories. According to the National Coalition Against Domestic Violence (NCADV), however, most violence toward women occurs in their homes and is perpetrated by people they know. Within their lifetimes, one in three women and one in four men have been (or will be) victims of physical violence by an intimate partner (Black et al. 2011). In addition, attention to sexual assault on campuses across the US involves critical analysis of how a culture of violence is shaped by various axes of power in these contexts. These discussions often challenge understanding of safe spaces. For example, many college campuses have emergency call centers strategically located in public spaces across the campus, while the vast majority of sexual assaults are committed in private spaces by a known assailant. Campuses have also implemented safe space or free speech zones. Consider the ideas and questions presented in Box 1.1 concerning the policies and use of these spaces on campuses that may be familiar to you.

Feminist geography examines how private spaces and the protection they are expected to afford become incorporated into public spaces to further

Box 1.1 Free speech, safe space, and campus climate

Campus communities are paying increasing attention to behavior, actions, and words that may be offensive or harmful to people who are members of their campus. This trend has raised awareness among students, faculty, staff, and administrators about how their actions and words may affect other people. For example, faculty use terms such as 'trigger warnings' before presenting material that may cause an emotional response among some students, and 'microaggression,' which refers to comments or a choice of words that some people perceive to be violent or offensive. These comments disproportionately impact underrepresented minorities, women, and LGBTQ students who feel marginalized by comments that depict stereotypes or discrimination.

The issue of free speech and safe space impacts classrooms, publications, other forms of communication, and the overall environment on campuses across the country. At the campus scale, some universities and colleges have spaces or areas on campus that are designated as free speech zones or areas where people can express their ideas without fear of repercussions. Many colleges instituted these areas of free speech during the Vietnam era, when some college campuses were sites of huge protests in the US.

Figure 1.3 Free speech zone at Iowa State University. Reproduced courtesy of Ann Oberhauser.

Activity

1. Do you have a free speech zone on your campus? If so, find out where it is located.
2. Who uses this space and for what purposes?
3. Are there competing groups or differences among issues that are represented in these zones? Do some people find the views offensive or harmful?
4. What organization or university body oversees the area and regulates who can and cannot use it?
5. How do these spaces reflect different perceptions of areas on campus and especially the importance of free expression?

understand social constructions of space. For example, as noted at the beginning of this chapter, veiling or body covering is often used as a method of signaling the privacy of the body in public space. Veiling can act as a form of what Hanna Papanek and Gail Minault (1982) label 'portable seclusion' or a mobile version of the privacy and protection provided by the home. In this example, place matters in understanding how veiling signals the privacy of the body in public spaces.

Additionally, public space is sometimes perceived by many in the majority group as open, accessible, and unregulated. Many public areas, however, have surveillance and policing that restricts free speech and mobility. Moreover, as Chapter 4 details, public spaces are often subject to certain norms of behavior. In certain cases, some people and their practices are seen to 'fit' while others are not. In sum, examining the way in which norms shape people's inclusion and exclusion in public and private space is a key area of focus for feminist geographers.

Scale and mobility in feminist geography

Feminist geography also contributes to the discipline of geography through its examination of the dynamic and contested aspects of **scale**, or the representation of areas such as the body, households, and the nation that are socially produced and thus potential sites of political struggle. The concept of scale has been fundamentally reframed through feminist geography's critical approach to understanding the situational and contested nature of the term. Rather than approaching scale as an absolute and static category, feminist geography argues that it is a category of power dynamics with contested and fluid characteristics. In their work on urban politics, David Delaney and Helga Leitner (1997) argue that our construction of scale impacts the 'ways in which we go about creating, revising, and living within a complex set of power relations' (94).

Feminist geographers have also theorized the body as a key scale of analysis that is entangled with other geographic scales. For example, Michael Brown's (2000) research on sexuality and space uses the term 'closet' as a spatial metaphor to examine power relations. Brown organizes his discussion according to scales, i.e., the body, urban space, the nation, and the world, to show how the concept of closet space is useful in examining these power relations. Feminist geography maintains that scales are mutually constituted and overlapping—not discrete forms of analysis (Marston et al. 2005). Similarly, feminist geographers reveal dynamic connections between the global (or more expansive) scales and the intimate (or confined) scales such as the body (Pratt and Rosner 2012).

Sara Smith's (2012) analysis of intimate geopolitics and reproductive bodies, for instance, examines the ways in which changing political structures in India have caused divisions between multi-religious communities. She argues that in Leh, Ladakh, in the state of Jammu and Kashmir in Northern India, Muslims and Buddhists had a significantly long history of intermarriage until political changes at the national scale filtered into this community, causing a ban on inter-religious marriages. This example illustrates how political changes at the national scale impact the most intimate of decisions, such as marriage and childbearing.

In studies related to the dynamic intersections of multiple scales, feminist geography has shown that mobility and migration are gendered processes. **Mobility** refers to the physical movement of people through either long-term relocation or short-term movement in everyday life. Mobility can also be a social process such as a change in socioeconomic status from lower to higher status. Gendered and racialized relations and access to resources in many societies influence the extent and frequency of mobility, from daily commutes to work to the flight of refugees. Palestinians traveling within Israel and the occupied Palestinian territories provide a clear example of the way in which intersectional identities can result in differences in mobility. Through the use of identity cards and travel permits, the mobility of Palestinians is determined based upon an individual's characteristics (such as their age, sex, occupation, and political involvement) and the characteristics of the community in which they live. These characteristics determine when, where, and for how long an individual is permitted to travel (Alatout 2009). Dowler's (1998) work in Northern Ireland reinforces the importance of gender for mobility in conflict areas, as she demonstrates that the association of women with the home can ironically mean increased mobility for them across borders in such militarized contexts. However, it is not only in highly regulated or conflict areas that mobility differences exist; rather, research on daily mobility patterns demonstrates that among women in the US, African American women experience the longest commutes, as a result of their reliance on public transportation and 'poor spatial accessibility to jobs' resulting from the racialized nature of historical and contemporary housing markets, labor markets, and transportation systems (Parks 2016, 295). In addition to a lack of transportation, experiences of 'policability' and feelings of surveillance by white neighbors contribute to limited mobility among Latinos as a whole, and Latina women in particular, in the context of the US (Maldonado et al. 2016).

Introduction and overview

As an important component of mobility, **migration** is also a key topic in feminist geography. As noted here, feminist perspectives on migration shed light on some of the driving forces and societal impacts of the permanent or temporary mobility of people. Women's migration has increased in recent decades as opportunities for 'care work' have expanded in developed regions and parts of the Global North. In turn, state and national politics often define the context and regulation of global migration and refugees. In Chapter 6, a focus on feminist geopolitics addresses the contested and highly fluid movement of people who voluntarily or forcibly leave their countries of origin for economic, political, and cultural reasons (Silvey 2004; Hyndman 2012). Gender differences are evident in labor migration, whereby men commonly move for certain forms of work in manufacturing, construction, agriculture, and extractive industries such as mining. In contrast, female migration is predominantly for work in the service sector in the form of domestic labor and childcare (Pratt 2005). In short, mobility and migration are deeply gendered processes.

Feminist understandings of the environment

The ways in which feminist geography is practiced and theorized also draw from geography's emphasis on **human–environment relations**. With strong influences from **ecofeminism**, a field of scholarship that established a link between the patriarchal oppression of women and the exploitation of the environment, **feminist political ecology (FPE)** has emerged as a key realm of feminist geography (Seager 1993; Carney 1996). FPE is a prime means of understanding the way in which gender and other kinds of difference are embedded in human–environment relations. As a critique of the larger subfield of political ecology, FPE responded to political ecology's scant attention to gendered power and knowledges in the realms of natural resource struggles, subsistence farming, climate change, livelihood struggles, and population control among other areas (Mollett 2010; Rocheleau et al. 1996; Sundberg 2004).

Research on environmental themes by feminist political ecologists Dianne Rocheleau, Barbara Thomas-Slayter, and Esther Wangari (1996), Sharlene Mollett and Caroline Faria (2013), and Andrea Nightingale (2011) examines the multiple aspects of dominance and control over nature in areas such as race, class, ethnicity, and gender. Thus, human–environment, or 'nature–society,' relations are closely entwined with contested and fluid gendered divisions of labor and power relations. Increased attention to interspecies relations has also enriched and expanded the scope of the overlapping spaces

Box 1.2 Contested views of development

The use of certain terms to describe economic development has been a continual process that attempts to categorize countries and people. The following overview provides definitions of some of these terms in the field of development studies.

First World, Second World, Third World: During the Cold War between the US and the former Soviet Union, 'First World' was the term used to identify the capitalist countries, 'Second World' identified the communist countries, and 'Third World' initially described those places that were beyond the capitalist/communist worlds (i.e., the countries that were aligned with neither the US nor the Soviet Union). At the end of World War II, 'Third World' became synonymous with 'developing countries.' In contrast, the term 'developed countries' came to signify industrialized and post-industrialized countries, which were economically 'developed,' while non-industrialized countries were labeled 'underdeveloped' or 'developing.'

Similar discursive constructions lie in such dichotomous terms as 'West'/'East' and 'Global North'/'Global South' respectively. For the former, the 'Western' world refers to the historical colonization by Western European powers and their influence in the East, such as Asia and Africa. While 'West' continues to be used to refer to Europe, the US, and Canada, it is also not geographically accurate as there are countries east of Europe that are more economically and ideologically similar to Europe, just as there are countries west of Europe (and south of the US) that are significantly different, economically and ideologically, from the US or Europe. Similarly, 'Global North' refers to countries in the northern hemisphere and 'Global South' to countries in the southern hemisphere.

In response to concerns related to these various concepts, the terms 'One-Third World' and 'Two-Thirds World' have been used to avoid the problems of inaccurately associating economic conditions with certain regions by calling attention to the economic and political differences between people across geographic space. 'One-Third World' identifies individuals and groups with more economic and political power and influence globally, compared with 'Two-Thirds World,' which identifies the majority of individuals on the planet without significant economic or political power.

More recently, terms such as 'the bottom billion' have been used to identify people living in extreme poverty. Similarly, Occupy Wall Street activists used the terms 'the 1%' to identify the economically privileged few and 'the 99%' for the vast majority of those at the other end of the economic hierarchy.

of animal geographies and feminist geography. Alice Hovorka (2015, 1), for example, argues for the need to 'explore and enhance feminist theoretical dialogue about, on and with animals in order to generate geographical knowledge that is comprehensive and reflective of the interrelatedness of all beings.' Such recent contributions have sparked interesting debates and new meanings of gender difference.

Finally, human–environment links to development and globalization are greatly enhanced by feminist geography. As outlined in Box 1.2, 'development' is a contested term that reflects colonial legacies and social categories with an emphasis on the relationship among and between systems of domination. Likewise, the explanation of how globalization works within and across borders is linked to dynamic gender identities and unequal gender roles in diverse contexts. Consider the impact of globalization in the production and consumption of goods or services, or the way in which globalization influences the cultural construction of the 'other' in song, clothing, or language; feminist approaches help us to understand distinct and diverse experiences and manifestations of globalization, gender, and place.

In sum, central themes and research interests within feminist geography highlight the way in which power relations are multiple and occur across space and time. Geographic processes such as human–environment relations, development, and globalization inform the contested and fluid aspects of our lives and society. How feminist geography arrived at this juncture is discussed in the following section.

Feminist geography: a genealogy of research on women, gender, and feminisms

While we open our book by highlighting the importance of intersectional thinking, such a priority in practice is relatively new. The field of feminist geography has evolved through many stages to better understand the relationships among social equality, power relations, and spatial dynamics. Early approaches to feminist geography addressed inequalities surrounding the role and status of women in society (Domosh and Seager 2001). This area of geography coincided with developments in the field of women's and gender studies, movements such as second-wave feminism, and increasing political pressure in many regions of the world to expand the rights of women. In geography, feminism drew from the belief that the category of 'woman' and women's experiences were not being studied within human geography. In research from the 1980s, feminist geographers Janice Monk and Susan Hanson (1982) raised questions about the implications of excluding 'half of the human in human geography' because of the relatively low numbers of women as geographers and subjects of study. In addition, Linda McDowell (1991) challenged the omission of analyses concerning women in urban space, while Janet Momsen and Janet Townsend (1987) addressed issues pertaining to women in international development. These and other foundational contributions suggested ways in which the discipline could theoretically and methodologically address the need to include issues relating to women and gender in geographic analyses.

Early studies of women and geography also examined how women's experiences and roles differed from men's and addressed alternative ways of 'counting' women (Domosh and Seager 2001). Analyses of issues that affected women, such as equal pay, job segregation, and maternity leave, were instrumental in pushing the boundaries of research on women in the workplace. Thus, this research meant to challenge masculinist ways of conducting research and analyzing socioeconomic patterns in what was labeled by Monk and Hanson (1982) as 'sexist biases in geographic research' (11). The marginalization of women and gender issues in the discipline involved the omission of these concepts, as well as the neglect of women's representation in the field. As Lynn Staeheli and Patricia Martin (2000) argue, 'the development of feminist geography has been closely linked to the position of women in the discipline. This position has been structured by the intellectual history of the discipline, the paucity of women in the discipline until recently, and a dominant mode of research' (137).

Starting in the 1970s, radical voices in the discipline of geography resonated with feminist ideas about exploitation and power relations. Structural approaches grounded in socialism and Marxist political economy examine class as an important category working alongside patriarchy as integral to the framework of capital accumulation. **Socialist feminists** argue that patriarchy and capitalism are forms of exploitation whereby working-class women face double discrimination. As such, rather than taking a **liberal feminist** approach that argues that women are not given their 'fair share' in the context of so-called neutral social systems—for instance, by not being paid equally to men in the capitalist economy—socialist and other radical feminists critique the systems themselves. In a seminal contribution to these debates, Sylvia Walby's (1990) *Theorizing Patriarchy* outlined six structures by which men dominate women in advanced industrial societies: households, waged work, the state, violence, sexuality, and cultural institutions (the media). Chapter 5, which focuses on the economy, expands on these multiple dimensions of

Introduction and overview

gender and economic activities within different geographic contexts. In addition, Deniz Kandiyoti (1988) developed the notion of patriarchal regimes, which refers to differentiated patriarchal structures that operate in many parts of the world. Her work has also been instrumental in demonstrating how women negotiate and bargain with patriarchy in individual contexts to protect their social and economic positions. This research underscores various aspects of women's agency within different patriarchal structures.

Another challenge to the unified subject and a deconstruction of hegemonic narratives is offered by **feminist post-structuralism**. Theorists in this school of thought challenge dominant and often masculinist and Western views of society and space. Instead, they offer an approach that considers why we are interested in particular objects of study, and that examines the systems of knowledge that produced these objects. This approach was influenced by philosophers such as Jacques Derrida, Michel Foucault, and Nancy Fraser, who focus on discourses or dominant narratives that marginalize people and points of view that are less visible or lack power in society. Post-structuralism thus brings attention to the process of 'othering' people and voices, which in turn dismisses marginal lived experiences and outlooks. J. K. Gibson-Graham (2006), Gillian Rose (1993), and Geraldine Pratt (1993) have advanced these ideas in feminist geography by challenging dominant narratives about capitalism and patriarchal relations of production. As discussed in Chapter 5, Gibson-Graham's (2006) work on alternative economic strategies provides a view of household and community-based livelihoods that are outside the realm of capitalist relations of production. Instead of dualisms or binary aspects of the social and spatial, post-structuralist feminism focuses on overlapping and integrated elements of social and spatial processes.

Many of these early approaches in feminist geography were grounded in Western experiences and perspectives on gender and women. Starting in the 1980s, voices from the Global South began to influence not only feminist geography, but also feminism as a whole. In particular, **post-colonial feminists** such as Chandra Mohanty (2003a), Gayatri Spivak (1999), and Alison Blunt and Gillian Rose (1994) have expanded critiques of Western feminism and dominant approaches to gender, women, and feminism. Colonial and imperialist manifestations of feminism used the idea of women's rights as a reason for invasion or intervention through colonialism, military invasion/occupation, and international economic development. Therefore, 'Third World' women were framed as oppressed by their own cultures and patriarchies in order for the occupiers to position themselves as 'saving' women, rather than socially engineering and exploiting populations, economies, and territories. Mohanty (1991) demonstrates that ahistorical approaches to patriarchy and women imply, 'the erasure of the history and effects of contemporary imperialism that underscores third world women's oppression' (34).

Post-colonial feminism draws from these historical contexts of women's status and locations in former colonial regions of the Global South, where intersections of class, ethnicity, religion, sexuality, gender, race, and/or caste shape their experiences and inform feminist approaches to difference and identity. For example, Structural Adjustment Programs and loans by the International Monetary Fund in many areas of the Global South exemplify neoliberal policies that have had longstanding impact on these countries' economies and women's and men's livelihoods. As a result, for feminists considering women's issues in such contexts, analyzing and critiquing the neo-colonial economic and political systems that support such institutions and policies is critical to understanding and addressing the intersectional nature of women's subordination. Grounded in this post-colonial perspective, **transnational feminism** further develops these ideas by focusing on the fluid connections and boundaries that define the cross-cultural integration of various feminisms. Feminist geographers Geraldine Pratt and Brenda Yeoh (2003), Cindi Katz (2004), and Amanda Lock Swarr and Richa Nagar (2010) interrogate political binaries (such as local/global, center/periphery, and Global North/Global South) in order to show how they inform each other and are interrelated, rather than distinct or separate categories. In this context, women's lives are connected via different sets of transnational relations and practices that cut across national boundaries and economic forces. Transnational feminism also focuses on praxis and ways to challenge dominant boundaries and structures that divide and marginalize women. Valentine Moghadam (2005), for instance, highlights transnational feminist networks (TFNs) as a means of providing alternative visions for both local and global causes. These networks represent ways to develop transgressive and emancipatory interventions and collaborations among women.

Additional critiques of white, Western-dominated approaches to feminism were based on many issues of discrimination and inequality experienced by women of color. **Critical race theory** (CRT), advanced in scholarly relation to critical legal studies, emerged in the early 1970s as a critique of both liberal and conservative notions that the law and its related institutions (namely, the courts and police) are the product of, and legitimized by, a set of neutral and objective rules. Rather, critical legal studies scholars argue that the law is entirely inseparable from the political processes that reproduce both

Introduction and overview

socioeconomic privilege and disadvantage. Critical race theorists extend this understanding to include how the law also reproduces and justifies racialized power and white supremacy. This theory has blended with other critical branches of geography such as Black geographies, transnational feminism, and post-colonial feminist geographies (to name a few).

Feminist geographers have over time increasingly engaged with CRT, such as in examining how space is made and re-made in both urban and rural contexts, and in critiquing the discipline of geography itself. For instance, feminist and environmental justice geographer, Laura Pulido (2000, 2006) highlights how in many ways white privilege and most poignantly white supremacy informs environment degradation, a spatial condition that people of color and their communities experience in disproportionate rates compared to white communities in the US. However, attention to race in geography continues to be marked by tensions. For instance, in her Past President's address at the Annual Meeting for the American Association of Geographers, Audrey Kobayashi (2014) acknowledges a long history of geographers' dismissal of racial studies and placement of racial justice outside the field of geography, assuming it belonged to other disciplines such as sociology and anthropology. This history of a lack of importance ascribed to scholarship on racialization and the theories that underpin such processes also included the devaluation of those scholars who have, all along, been fighting to make racial power visible in geography and the social sciences. She writes, 'the concept of race has a benighted biography among those who have created our discipline. I take the opportunity of this address to delve into the race concept and a situated series of enactments and interactions among geographers who have shaped its evolution' (1101). Kobayashi's lifetime anti-racist feminist geographic work is joined by that of a growing group of scholars focusing on race, whiteness, environmental racism, race and political ecology, gentrification, urban political ecology, imperialist and historical geographies, and Black geographies (to name a few), and has, as we will see in future chapters, shaped the scholarly directions within feminist geography.

In recent decades, **queer geographies** and related themes concerning sexuality and space have further expanded the scope and critical understanding of how power relations are constructed in different social and spatial contexts. Drawing from work in post-structuralist and post-colonial feminism, queer geographers such as Catherine Nash (2008), Natalie Oswin (2008), Lynda Johnston and Robyn Longhurst (2010), Petra Doan (2010), and David Bell and Gill Valentine (1995) examine diverse spaces and identities that transgress

Figure 1.4 Trans March during the Gay Pride Parade in San Francisco, California. Trans activists increasingly use space to work for recognition of their identity and legitimacy. Reproduced courtesy of Quinn Dombrowski.

and challenge heteronormative societal relations. By implementing measures to disrupt these spaces, this approach aims to create non-hegemonic and inclusive spaces. Work by Doan (2010), for example, reveals the 'tyranny of gendered spaces' for transgendered people where gender normativity prevails in various public, semi-private, and private locations (636). In instances such as these, queer geographies represent a challenge to conventional ideas about power and voice that reflect fragmented and alternative approaches to relations of power (Figure 1.4).

It is important to note that the organization and structure of feminist geography does not form a linear or cohesive trajectory over time. Instead, advances in feminist geography have established debates and posed new questions in a way that represents overlapping, fluid, and sometimes contested approaches. The theoretical frameworks and empirical examples addressed here borrow from other approaches in geography and within critical social science, as they build on dynamic and shifting understandings of power and identity.

Finally, this book highlights how feminist geography is grounded in activism. A useful concept in this discussion is **praxis** or developing a way to practice feminism that is informed by a conceptual understanding of gendered power relations. According to Pamela Moss and Karen Falconer Al-Hindi (2008), feminism, and feminist geography more specifically, analyze and challenge inequalities and uneven power relations that privilege certain groups over others,

Introduction and overview

and develop strategies to advance progressive movements in society. Our discussion of feminist geography provides in-depth analyses and explanations of broad trends and everyday experiences that focus on themes such as the mutually constitutive nature of space and gender, power relations, difference, and praxis. Much of this discussion is embedded in the ability to apply praxis to feminist research through attention to research methodologies. These approaches include ethnography, participation, and engagement with people and communities. They also involve attention to positionality and reflexivity for the researcher. For example, Nagar (2002) refers to the need to decolonize research through collaboration and work that benefits all parties who are involved in the research endeavor. Praxis involves feminist pedagogy that orients teaching and learning to critical reflection on feminist themes and practice.

About the book

Feminist Spaces is a co-authored text for which each of us drew from our different research projects, activism, and experiences to write our respective chapters. As a result, this book contains multiple voices and approaches. We think of this—inspired by intersectional thinking—as an epistemological move against uniformity or positivism and toward diverse and nuanced synopses. In keeping with the theme of difference, we write this textbook as individual scholars drawing from each of our own research strengths and as a collective of feminist geographers who are committed to training future feminist geographers in a refreshing and critical way. This book is intended to help students understand global phenomena as well as their own households and communities. In the final pages of this section, we briefly outline the chapters and organization of the book.

In Chapter 2, Jennifer Fluri expands on the concepts of scale, space, and gender as they relate to the body. Feminists have long argued that bodies are socially constructed and inscribed entities reflecting and influencing societal values and power relations. Embodied aspects of how we dress, eat, and move through space represent the bodily performances that feminist scholars such as Judith Butler have examined in their work. Also important to this research is the growing critique of the gender binary within feminist, queer, and transgendered geographies.

In Chapter 3, Risa Whitson links the social meaning and practice of culture to feminist geography. This chapter redefines normative and hegemonic approaches to culture and gender by examining the home and sites of consumption as critical to the expression of gendered identities across cultural contexts. The discussion also analyzes geographic representations of social identity in the media and popular culture. Social media has transformed human interaction, including paving the way for feminist organizing and activism in the context of anti-globalization campaigns, grassroots political movements, and combating sex trafficking.

In Chapter 4, Whitson continues feminist geography's long tradition of analyzing women's roles and activities in urban spaces. The city is often the backdrop for analyses of gender norms and relations in public space, the workplace, and the home. This chapter examines how considering the right to the city from a gendered perspective offers new ways of thinking about social identity, public spaces, and the built environment. This perspective involves a focus on how gender and difference intersect both with experiences of public space and with the growth of the neoliberal city, examining the issues of mobility, fear, social reproduction, and gentrification.

In Chapter 5, Ann Oberhauser examines how the definition of work is explored from a feminist perspective, alongside related issues concerning income, the 'care chain' of labor, gendered divisions of labor, and the informal sector. This discussion also considers transnational migration, along with the displacement of families and individuals. Finally, feminist perspectives on globalization demonstrate how resistance to hegemonic neoliberal economic policy takes shape through a variety of struggles against exploitative industrial and labor practices and in the creation of alternative economic strategies.

In Chapter 6, Fluri explores gendered geopolitical concepts of territory, nationalism, and the state. She demonstrates that the history of nation building and control over territory often assigns different roles for women and men in defending, constructing, and maintaining the nation and state. These and other feminist approaches to geopolitics and militarization as a method of gendered and racialized control and domination form the basis for this discussion.

In Chapter 7, Sharlene Mollett familiarizes readers with the field of feminist political ecology (FPE). Readers are introduced to how feminist political ecologists employ various concepts and meanings as a critique of international development policy and practice in the Global South. Throughout the chapter, she demonstrates, through various examples and case studies, how FPE attends to intersectional power relations whereby race, religion, emotions, caste, and class are entangled with gender. These intersections are not simply forms of oppression but can be sources of resistance. Using FPE as a critique of international development, she illustrates how the contributions of FPE hold transnational sensibilities, blurring the lines between the Global South and the Global North.

In the final chapter, we provide an overview of the chapters and reflect on how insights from feminist geography align with other growing subfields in human geography such as indigenous, Black, and LGBTQ geographies. Feminist geography and these related fields benefit from intersectional understandings of power that challenge dominant notions of knowledge production. The struggles that advance these contributions also create solidarity and alliances that are built on a commitment to diverse, plural, and inclusive scholarship, praxis, and activism.

As we opened this chapter with insights from SlutWalk, we write this book knowing that feminist solidarities cannot be presumed or taken for granted. This book traces and reflects the contested nature of various influences in feminist geography that offer a different way of seeing our world. Overall, feminist geographies provide a platform to analyze and interpret the privileges, disparities, and inequalities that constitute a myriad of gendered social relations and practices in society. Our hope is that both students and instructors envision the possibilities of feminist geographic thought and praxis that hold the promise of empowerment and progressive social change.

Recommended reading

Domosh, Mona, and Joni Seager. 2001. *Putting Women in Place: Feminist Geographers Make Sense of the World*. New York: Guilford Press.

Gibson-Graham, J. K. 2006. *The End of Capitalism (As We Knew It): A Feminist Critique of Political Economy*. Minneapolis, MN: University of Minnesota Press.

Katz, Cindi. 2004. *Growing Up Global: Economic Restructuring and Children's Everyday Lives*. Minneapolis, MN: University of Minnesota Press.

McDowell, Linda. 2013. *Working Lives: Gender, Migration and Employment in Britain, 1945–2007*. Oxford: Wiley-Blackwell.

Mohanty, Chandra Talpade. 2003. *Feminism without Borders: Decolonizing Theory, Practicing Solidarity*. Durham, NC: Duke University Press.

Monk, Janice, and Susan Hanson. 1982. 'On Not Excluding Half of the Human in Human Geography.' *The Professional Geographer* 34 (1): 11–23. doi:10.1111/j.0033-0124.1982.00011.x.

Moss, Pamela, and Karen Falconer Al-Hindi. 2008. *Feminisms in Geography: Rethinking Space, Place, and Knowledges.* Lanham, MD: Rowman & Littlefield.

Oswin, Natalie. 2008. 'Critical Geographies and the Uses of Sexuality: Deconstructing Queer Space.' *Progress in Human Geography* 32 (1): 89–103.

Rocheleau, Dianne E., Barbara P. Thomas-Slayter, and Esther Wangari, eds. 1996. *Feminist Political Ecology: Global Issues and Local Experiences.* International Studies of Women and Place. New York: Routledge.

Rose, Gillian. 1993. *Feminism and Geography: The Limits of Geographical Knowledge.* Minneapolis, MN: University of Minnesota Press.

Smith, Sara. 2012. 'Intimate Geopolitics: Religion, Marriage, and Reproductive Bodies in Leh, Ladakh.' *Annals of the Association of American Geographers* 102 (6): 1511–28. doi:10.1080/00045608.2012.660391.

Swarr, Amanda Lock, and Richa Nagar, eds. 2010. *Critical Transnational Feminist Praxis.* Albany, NY: State University of New York (SUNY) Press.

Chapter 2

THE BODY, PERFORMANCE, AND SPACE

Jennifer L. Fluri

Chapter 2

Fluri

Introduction

The body is a surface upon which various forms of social, cultural, and political meaning are inscribed. Our bodies are marked and we mark our bodies in various ways. The clothes we wear, the melanin in our skin, the color of our hair and eyes, our ability or inability to afford certain corporeal care products: all communicate certain aspects of who we are and how others see and perceive us. The body's location in certain places provides another form of communication. For example, you may dress differently at home, work, school, or other locations. Certain groups of individuals may be categorized based on their corporeality and subsequently associated with a particular geographic place. The ease or difficulties we experience while traversing through different spaces illustrate other aspects of what we call **corporeal geographies**. This chapter provides an overview of the social, political, and economic aspects of corporeal geographies, beginning with a discussion of how gender is conceptualized.

In this chapter, the social construction of gender and performativity are defined along with the concept of **subjectivity**. These explanations are followed by an exploration of the interchanging relationships between bodies and geographic spaces. **Corporeal markers** such as gender, race, class, disability, sexuality, and dress have been infused with social, political, and economic meanings that are spatially organized. The spatial organization of corporeal markers leads into a discussion of cultural bodies (Valentine 2014). A cultural body refers to the ways in which acts and actions of individual bodies become representative of an entire group because of their corporeal similarity to that group. Dress is another corporeal marker and way of communicating one's identity or association with a particular group or ideology. Dress has been used as a method of controlling bodies in certain spaces by governments and other powerful institutions. This chapter concludes by discussing various forms of corporeal control (and resistance to control) in different locations and contexts.

Gender and the body

Gender is an expression of the social roles, norms, and expectations that are mapped onto one's biological body. Nonetheless, biological sexual differences between women and men are not firmly dichotomous. The intersexed—people born with (or who develop) indiscriminate genitalia—exemplify the lack of a strict biological binary between male and female. Gender differences between men and women are often produced or constructed through social relations. Feminist philosopher Judith Butler analyzed the ways in which gender manifests onto our bodies through the concept of **performativity**. Butler's seminal work in *Gender Trouble* (1990) and *Bodies that Matter* (1993) identifies gender as the act of doing (i.e., repetitive acts associated with a particular gender) rather than being (biologically determined by genitalia). She argues that learned social behaviors about gender roles become viewed as normal due to the repetitive performativity of specific behaviors over time. Butler further argues that what we assume to be an internalized embodiment of gender is truly an assemblage of continuous acts/actions. These acts manifest onto the body through various forms of modification, stylizations, and behaviors.

Robyn Longhurst (2008) draws on Butler's theories to examine the multiple ways in which pregnant bodies and mothering are influenced by, and have an influence on, cultural and social processes. By examining maternal bodies through a sociocultural lens, Longhurst challenges the association of women's maternal bodies as a 'natural' expression of femininity. She argues that just as men can be, but are often overlooked as, nurturers, not all women feel a 'natural' or inherent desire to give birth, or raise and care for children. Her research shows how pregnant bodies are monitored and controlled in societies through various forms of surveillance and discipline about the places maternal bodies should and should not traverse. For example, she discusses pregnant women's fears about public censure for certain behaviors, such as drinking wine or going to a bar or nightclub. Other forms of policing maternal bodies include restricting when and where women can breastfeed in public spaces. Maternal bodies can be political spaces for governments or political groups seeking to control (or resist the control of) biological and social reproduction. The politics of maternal bodies and motherhood is discussed in more detail in Chapter 6. The geographies of maternal bodies highlight the manifestation of dominant social, cultural, and political discourses.

Dominant social or political discourses and expectations about gender establish norms that require repetitive actions for them to be effectively performed onto bodies. For example, dominant expectations about gender can occur in seemingly banal ways, such as when hospitals assign colored caps to newborn bodies based on their sex, such as pink for girls and blue for boys. The displacement of dominant gender roles, according to Butler, requires a 'slippage' within the process of repetitive performativity. This 'slippage' refers to alternative ways of being male or female, which resist or challenge prevailing representations or expectations of gender roles or norms. Therefore, one method of questioning conventional gender roles has often been to alter or challenge prescribed norms. 'Queering' is one method of embracing gender 'slippages' to challenge the status quo.

Several geographers have incorporated Butler's philosophical perspectives into their research on corporeal geographies. Geographic analyses of gender performativity show that the doing of gender and other forms of identity must incorporate an understanding of a person's lived history, their interlinked relationships with others, and their connection to, and interrelationship within, a particular place or moment in time (Nelson 1999). In other words, gender as it intersects with other social categories (such as race, class, and sexuality) has been shaped through various assemblages associated with places, people, situations, and experiences.

Subjectivity is another philosophical concept taken up by feminist geographers. Subjectivity refers to the ways in which experience shapes understanding. Dominant forms of white masculinity have developed spatial imaginaries that are implicitly violent. Therefore, feminists have tried to think differently, in ways that are more inclusive, and to promote 'positivity of otherness' to resist racism, sexism, and other forms of violent oppression (Nagar 2014, 150). Feminist inquiries have therefore questioned the powerful assumptions made about individuals based on the gender, color, shape, phenotype, or ability of their bodies. The inherent privileges that certain bodies experience based on the way that they look must be challenged, along with the prejudices and exclusions that other bodies experience based on appearance. Nagar (2014) suggests that in order to effectively challenge these subjective inequalities we should practice **radical vulnerability**. This 'requires all members of an alliance to open ourselves—intellectually and emotionally—to critique in ways that can allow us to be interrogated and assessed by one another. . .' (15). By allowing ourselves to be vulnerable—rather than holding onto preconceived notions of our superiority, inferiority, strengths, or weaknesses—we can open up a space to challenge the categories that separate us in order to have a chance to listen, speak, interact, and learn from one another.

Feminist geographers use the concept of **intersectionality** (discussed in Chapter 1) in order to both understand identities as multiple and how they are shaped through power relations, structural violence, and unequal access to social, political, and economic rights. Even a cursory evaluation of gender trends over time and across space helps to illustrate how gender norms are formed to meet different and changing social conventions or political agendas. Box 2.1 explores changing gender norms in the US during and after World

The body, performance, and space

Chapter 2

War II and provides an activity for thinking about what cultural icons, such as Rosie the Riveter, endure today.

Box 2.1 Examples of changing gender norms

In the US during World War II, men were expected to join the military and leave home as part of the war effort. Correspondingly, women were expected to support the war effort by working in factories to replace the men who left for war. In an effort to normalize these changing gender norms, welding was compared to knitting. The infamous icon Rosie the Riveter has been associated with the 'We Can Do It' poster (Figure 2.1). This poster depicted a white woman (Geraldine Doyle) wearing a head covering typical of factory workers, with painted eyebrows, eyelashes, rosy cheeks, and lipstick, while holding up her arm to show her muscles, along with the caption 'We Can Do It.' This poster attempted to encourage women to become factory workers and active participants in producing weapons at home for the US war effort internationally. This poster, along with others of this time period, simultaneously highlighted white femininity.

Women maintaining their femininity was seen as an expected part of their wartime duties, as evidenced by cosmetic advertisements during this era. The US War Production Board kept cosmetics off its list of rationed items for restricted wartime industries during World War II. The changing gender norms that encouraged women to work outside the home were temporary,

Figure 2.1 'We Can Do It!' poster for Westinghouse. The image is closely associated with Rosie the Riveter, although not a depiction of the cultural icon itself. Pictured: Geraldine Doyle (1924–2010), at age 17. Date: 1942.

Source: jpeg file retrieved from commons.wikimedia. org. Author: J. Howard Miller, artist employed by Westinghouse, poster used by the War Production Coordinating Committee. This work is in the public domain because it was published in the US between 1923 and 1977 and without a copyright notice.

29

though, as women were expected to leave their jobs and return to their homes at the end of the war. Women's return to the home would allow male veterans in the US to regain employment. Despite the temporary shift in labor-based gender roles, racial segregation dominated; both within the military and in many public spaces within the US, where Jim Crow laws segregated spaces based on race. African American men were expected to serve in the military and African American women in factories or as domestic servants in white homes, while their work and service was undervalued due to prevailing social and political power structures that reinforced racial segregation and vast inequalities based on race and gender.

When white women were expected to return to the home in the post-war era, the dominant discourses and representations of gender within the US were expressed through white middle-class suburbia: idealized through expectations of men leaving the home to work, and women caring for the home and children. This idealized picture of gender roles and relations was, however, limited to the white middle and upper classes, i.e., heterosexual, two-parent couples with children.

In this example, powerful organizations such as the US government sought to change distinct gender norms in an effort to support military action in World War II. After the war, political and economic expectations demanded a reordering of gender norms that were infused with similar racial, class, and spatial divisions. Interestingly, the 'We Can Do It' poster has endured beyond its use as a wartime propaganda poster. For example, liberal feminist activists in the 1970s reclaimed this image to rally support for women's rights activism and gender-pay equity.

Activity

1. Search images on the internet for the 'We Can Do It' poster.
2. Identify the various ways in which this iconic image has been reused by various individuals, organizations, and groups.
3. Of these images, which do you find the most compelling?
4. What do you think makes this image so enduring within US iconography?

Bodies as geographic spaces

Corporeal inscription represents the ways in which we shape our identities onto the body. The corporeal geographies of inscription manifest through different forms of bodily markers. There are some markers that we cannot easily change, such as skin color, facial structures, height, and shape. However, changing the body's biological features *can* occur through the use of cosmetics and surgical technologies. Cosmetic surgery has become a growing business in various parts

The body, performance, and space

of the world, as well as becoming a facet of medical tourism. There are distinct geographies associated with cosmetic surgery tourism. Advertisements for cosmetic surgery connect these corporeal augmentations to personal 'improvement' or 'success.' Similarly, the locations for this type of medical tourism are also sold as spaces of beauty, hope, and care. In this way, the places where cosmetic surgeries will take place are marketed as beautiful tourist sites to entice potential customer–patients to consume both the landscape and elective corporeal enhancement surgeries (Holliday et al. 2015).

Skin color and phenotype are corporeal markers that have become methods of categorizing individuals into associated groups. In a similar way to gender, race has been used as an identity marker, which has been layered with social, economic, and political meaning. Although there are no biological determinants associated with ability, or lack thereof, based solely on race or gender, racism and sexism remain rampant in many societies around the globe. Gender- and race-based categorizations continue to be used as divisive social and political hierarchies despite calls for gender and racial equality in civil and women's rights activism and human rights discourses. Geographers have examined the way in which these categorizations manifest into spatial access, marginalization, or exclusion.

Race and gender are spatially organized bodily markers. This means that race and gender are visible both on the body and by way of different geographies such as the workplace, the home, the city, the suburbs, etc. Understanding place-based differences allows us to examine how corporeally based inequalities are organized geographically (McKittrick and Peake 2005). For example, contemporary forms of racism and sexism can be traced to ideologies conceptualized during Western European colonization. White Western Europeans' racist treatment and abuse of people in Asia and Africa continues to reverberate through various forms of ongoing racism: the association of privilege and power with whiteness resonates today in many countries within Asia and Africa through various mechanisms that privilege lighter skin.

Cosmetic companies have manipulated and capitalized on the privilege associated with lighter skin by producing, advertising, and selling skin-whitening products. For example, advertisements for skin-lightening cosmetics in India reproduce historically based hierarchies of gender, race, class, and caste while claiming to 'liberate' Indian women and men from the constraints placed on them by these same hierarchies (Parameswaran and Cardoza 2009). In a similar way to other forms of advertising, these commercials reinforce existing social stigmas about race, gender, class, and caste in order to seduce consumers into believing that these products will help them to overcome corporeal limitations generated from racist and sexist power hierarchies (Oza 2006). These products also provide mechanisms for individuals to succumb to the expectations of racist and sexist subjugation, rather than challenging this dominant mode of thinking.

We can see similar trends in advertising within North America, where advertisements attempt to seduce consumers into buying products to meet a corporeal ideal that is often unattainable because the images in these advertisements have been significantly manipulated and rely on a narrow, standardized conception of beauty. For example, in 2011, GlobalDemocracy.com demanded mandatory disclaimers be identified on all manipulated images. (For an example, see the video 'Photoshop body evolution,' which shows the Photoshop transformation of a woman's body: www.huffingtonpost.com/2013/10/30/photoshop-body-evolution_n_4170021.html). 'Ideal' bodies have been socially and politically constructed based on the expectations of different organizations, groups, or governments. Advertisers' goal is to help their clients sell products. In other cases, ideal bodies have been put forth to represent a particular group or ideology. Representations of bodies are integral to the ways in which bodies are placed into different social and political categories. Bodies that do not meet the standards of the ideal body have been subsequently identified or 'coded' based on a series of assumptions. These assumptions are based on social and political ideologies that are then identified with, or mapped onto, different bodies.

Coding the body

Examining how cultural assumption is placed onto bodies provides a method of understanding the racism inherent in dividing people based on corporeal differences. Racist categorizations do not simply describe differences between bodies, but rather attempt to suggest that there are immutable biological divisions between people. These categorizations are further reinforced through relative comparisons to other bodies. Geographic analyses help us to see how corporeal differences are organized spatially and determined by social, political, and economic factors and influences, rather than being biologically determined.

Spatial analysis of bodies illustrates how power regimes have situated whiteness and masculinity in positions of privilege, making them seem like an unmarked category. Whiteness and masculinity are therefore constructed as a repudiation of non-whites and women as symbols of difference (Puwar 2004, 142). Racialized bodies exist within regimes of 'super-surveillance' where minor mistakes are put under a microscope, more often noticed by those in power or authority, and punished with greater severity (Puwar 2004, 143). Idealized white masculinity has been maintained by ensuring its distance from femininity and non-white bodies. This unequal treatment resonates among alleged criminals in the US, including the deaths of unarmed Black men and women by white police officers, which has prompted Black Lives Matter activism:

> Black Lives Matter is an ideological and political intervention in a world where Black lives are systematically and intentionally targeted for demise. It is an affirmation of Black folks' contributions to this society, our humanity, and our resilience in the face of deadly oppression.
> (blacklivesmatter.com/guiding-principles)

Attributing certain behaviors to one's corporeal identity is often referred to as 'coding.' **Code switching** is a term used to identify when individuals behave in ways that run counter to the assumed codes associated with their racial, gender, sexual, and class identity. When individuals code switch, such as women who perform their gender in masculine ways, or men who perform their gender in ways that are read as feminine, a slippage or destabilization of dominant gender norms can occur.

There are also people who perform their gender in ways that run counter to prescribed norms of sexuality. Queer and trans sexualities help to destabilize the strict bifurcation of sex (female/male) and gender (femininity/masculinity) (Browne 2004). Geographies of sexualities challenge conventional understandings of our sexual bodies, while explicating the ways in which our sexual bodies are also demarcated by other aspects of corporeality, such as race and class. Box 2.2 uses the case of Bruce/Caitlyn Jenner to discuss transgendered bodies as exemplifying flexible gender categories, while dominant social conventions simultaneously attempt to resituate transgendered bodies into fixed, rather than flexible, gender categories.

Gender, race, sexuality, and class codes have been inscribed onto bodies for the purposes of representing countries or national ideals and ideologies. In many countries, national affiliations are associated with a citizen's birth-location, parents' heritage, and/or association with a particular ethnic or racial group. Women are more often associated with biological reproduction and the subsequent care of children and social reproduction of households. Therefore, women have been situated as a repository of national and cultural identities

Box 2.2 Transgender transformations attempting to queer gender boundaries

Transgender bodies challenge the demarcation of gender roles into specific binary categories based partly on the visual representation of one's body. When Olympic athlete Bruce Jenner transitioned to Caitlyn Jenner, she took on a feminine appearance. The publicity and debate surrounding Jenner centered on her transition from a former representation as a physically masculine Olympic athlete, to that of a hyper-feminine and sexualized body as photographically presented in *Vanity Fair* magazine: (www.vanity-fair.com/hollywood/2015/06/caitlyn-jenner-bruce-cover-annie-leibovitz). The representation of her body relied on feminine stylish tropes. Stylist Jessica Diehl carefully selected the clothing worn by Jenner, and renowned photographer Annie Leibovitz took the photographs displayed in *Vanity Fair*. This transformation illustrated a bodily transition from masculine athlete to feminine cover girl, which, rather than challenge prescribed gender codes, reinforced them in a way that resituated masculinity and femininity into distinct (rather than fluid) categories.

Transgender bodies *can* suggest a fluid experience, flexible gender performances, and a slippage or challenge to conventional gender binaries. However, in this public representation of Jenner's transition/transformation, her 'new' gender was expressed through a particularly narrow lens of femininity. Rather than challenge the binary categories of masculine and feminine, this representation served to reinforce the borders that define a body as masculine *or* feminine instead of allowing a body to express various aspects of femininity *and* masculinity simultaneously. Thus, while transgendered bodies can (and often do) challenge socially constructed gender norms, this example of Jenner's transgendered transformation resituated gender into distinct categories. A more malleable way of expressing, doing, and being without strict gender boundaries, expectations, or codes clearly continues to produce social anxieties about how to 'do gender.' Therefore, this presentation of Jenner's trans-body relied on existing tropes, rather than presenting a multiple and flexible expression of gender.

and (at times) symbolic boundary markers of these identities (see Chapter 6). National beauty pageants are one example of a particular and idealized aspect of femininity within a nation.

The Miss America pageant began in the US shortly after the 19th Amendment passed, giving women the right to vote. Businessmen in Atlantic

The body, performance, and space

City, NJ, conceptualized the pageant as a scheme for attracting male consumers to return to this beach-resort town after the Labor Day holiday. Over time, this pageant became a national phenomenon and a way of expressing the 'ideal' young, female, corporeal representative of the US. This idealized woman has generally been white or light-skinned, and the pageant stipulates that she must be single and never have given birth. Various companies, from swimsuit and clothing manufacturers, to cosmetic companies, have sponsored this pageant. Several sponsors took the concept global, creating spin-off pageants, such as Miss World, Miss Universe, and Miss Earth. Caroline Faria's (2014) research on the beauty industry in South Sudan shows how fashion and beauty are interconnected with nation-building. For instance, her work shows local concerns about, or fears of, new and foreign styles and commodities associated with changing economic structures, as well as cultural shifts associated with returning migrants from the Sudanese diaspora. In many respects, embodying nationalism through fashion and style illustrates various aspects of corporeal geographies through gendered, raced, and sexed bodies.

Mr Universe pageants are exhibitions of sculpted bodies that predominantly emphasize idealized, hyper-muscular representations of masculine corporeality. The body modifications of both male and female bodybuilders express corporeal discipline and physical strength. Male bodybuilders often meet idealized stereotypes of masculine strength, while female bodybuilders disrupt conventional expectations of the public display of the female body (Johnston 1996). Lynda Johnston's research on female bodybuilders discusses the ways in which bodies and environments are mutually constituted. She identifies the gym as a space that confirms feminine and masculine bodily stereotypes, and specifically how different spaces in the gym become associated with masculine (free weights) or feminine (aerobics and circuit training) bodies. In her work, the colors of the room were also a way of coding spaces into male/masculine and female/feminine binaries. Her interviews with female bodybuilders showed how they 'transgressed hegemonic constructions of femininity' (333). The female bodybuilder exemplifies a 'slippage,' or challenge, to conventional notions of feminine corporeality by resisting the dominant discourses about gender, sex, and sexuality. Derek McCormack's (1999) research on fitness reveals how continuous messages about the ideal body type are promulgated by a multi-billion-dollar industry that manufactures a range of commodities associated with developing or maintaining a certain corporeal shape. He identifies fitness as a type of 'body-shopping,' where the beauty/

health industries construct advertisements that present an idealized and unrealistic image of the body in order to manipulate corporeal insecurities and encourage the consumption of body modification products.

As these examples have detailed, the concept of the 'ideal' body is socially, politically, or economically constructed for a particular purpose. By situating ideal bodies into narrow or unattainable categories, a variety of power regimes, from governments to corporations, exclude certain bodies or sell products and services that attempt to alter bodies to meet these problematic ideals. Therefore, bodies that fall outside of the manufactured 'ideal' have experienced different forms of exclusion, marginalization, or oppression. Abled bodies are another way in which idealized conceptualizations of 'fit, healthy, and beautiful' have been both reinforced and challenged.

Movement and (dis)ablism

In an effort to challenge the inherent divisions between abled and disabled bodies, Nancy Hansen and Chris Philo (2007) refocus geographies of the body on 'the normality of different bodies doing things differently' (493): 'We must force ourselves to rethink the ways in which disabled people – in fact, all people, disabled or non-disabled – occupy worldly spaces of all kinds. Disabled people themselves can (and should) be consulted for fresh perspectives on how to "map" this thoroughly spatialised bodily terrain' (501). Some aspects of disability studies within geography have focused on improving the accessibility of the socio-spatial environment, such as engineering spaces to be wheelchair accessible. Brendan Gleeson (1996, 1999) and Vera Chouinard (1999) examined individual experiences of disabled people in relation to mobility and access to spaces. Their research discussed the ways in which disability can be a spatially and socially constructed form of exclusion and oppression. They argued that disabilities have been socially and politically constructed to exclude individuals from certain spaces, subsequently leading to spatial and social marginalization. These studies incorporated the experiences of impairment from multiple perspectives and locations. Disability studies within geography further question the idealization of able-ism. This includes challenging the physical construction of built environments that remain exclusionary, particularly when they are built based on a narrowly defined conception of abled bodies. Abled bodies remain vulnerable to change, injury, and decreased able-bodied-ness across one's life cycle. An individual's type of disability (i.e., blind, deaf, physically disabled, etc.), must also be incorporated

The body, performance, and space

into our spatial understanding of disability because it is often a major aspect of how one experiences different spaces and places.

Geographies of disabilities have focused on experiences, rather than relying on objective measures of suffering, pain, or impairment. The experiences of impairment should also be considered within the social conditions of everyday life. For example, sociopolitical situations such as war, displacement, poverty, or malnutrition can cause or intensify a physical or mental condition. Technological enhancements to the body, in an effort to attend to disabilities, are another aspect of research within geography. For example, the technological advancements for prosthetic devices provide corporeal modifications in order to overcome the challenge of both 'natural' and built environments. Advancements in bionic prosthetics are re-engineering bodies to enable amputees to access *any* environment. Interestingly, advancements in high-tech prosthetic devices in the US have been significantly funded by the US military, due to the extensive number of servicemen and women returning with amputated limbs from the wars in Afghanistan and Iraq. Due to the high cost of bionic prosthetics, the ability to purchase these devices remains limited for many amputees. Corporeal modification to abled and disabled bodies, similar to other corporeal modifications, is accessible based on one's economic able-ness or ability to afford these modifications. This resonates with other discussions of how corporeal markers can include or exclude, provide or limit one's access to space, and be a marker of privilege or marginalization. Thus, the costliness of high-tech corporeal modifications (while providing amazing solutions for individuals) reproduces modes of exclusion, since they are not readily available or economically accessible to most individuals in need of these devices. The diversity of disabled bodies illustrates the importance of incorporating the experiences of individuals in an effort to understand the relationship between bodies and geographic space. The intersectional aspects of gender, race, sexuality, and class are integral to feminist geographic analyses of disability. Thus, the experiences of disabled bodies differ significantly because of the complexities of intersecting influences of culture, society, economics, and politics. Disabled bodies, beauty, and high-tech prosthetics are further explored in Box 2.3.

Box 2.3 Miss Landmine and Aimee Mullins

The Miss Landmine pageant exemplifies the use of aesthetics and visual performance to resituate the disabled body as beautiful. The pageant's organizers sought to draw public attention to the long-term problem of continued landmine explosions years after the end of conflict. Morten Traavik organized two Miss Landmine competitions: Miss Landmine Angola in 2008 and Miss Landmine Cambodia in 2009. As the privileged white male organizer, Traavik designed the concept and raised funds for this project. These pageants provided clothes and prizes to all contestants, and the winner received a state-of-the-art titanium leg, custom-designed in Norway, which costs an equivalent of US $15,000.

This pageant illustrates both national and transnational conceptualizations of corporeality, gendered aesthetics, and representation. The conceptualization of these pageants sought to challenge conventional expectations of beauty, while simultaneously relying on conventional aspects of feminine, heterosexual performativity to highlight the issue of landmines and their corporeal abuses. These competitions objectified female bodies while simultaneously critiquing the strict boundaries placed on women's bodies in conventional pageants. For example, many national and international beauty pageants require contestants to meet quantifiable bodily measures such as specific height, weight, and size specifications. Bodily specifications were not required in the Miss Landmine pageant (beyond having been injured by a landmine explosion).

The public representations of contestants' bodies as aesthetically pleasing through the performative aspects of pageantry garnered several paradoxical responses (as captured in the documentaries about, and news coverage of, these events), from pity and concern for the contestants to seeing their participation as empowerment or self-actualizing. The organizers further positioned women's bodies (on websites and in print publications) as a space for highlighting the physical costs of war against the actual costs of the munitions that caused their injuries. For example, in the contestant catalogue for the Cambodian pageant, images of each woman dressed in an American Apparel outfit were shown with text that identified the costs of the dress, shoes, and accessories, contrasted with the cost of the munitions that caused her injury. The contestant information further included the location of her home, where the landmine explosions occurred that claimed part of her body, and the landmine's country of origin/manufacture. Thus, the contestants' bodies provided a visually striking method of illustrating and detailing the costs and geographies associated with the manufacture and use of landmines.

These pageants used performance, image, and film as a method of 'challenging' conventional beauty expectations for pageant competition and the viewer's ability to see the disabled and amputated war body as an aesthetically pleasing political space. The winner of the Miss Landmine pageant received a titanium prosthetic limb that remained economically unavailable to her and the other contestants. The Miss Landmine website positioned 'beauty' under a discursive human rights umbrella, as suggested by the tag line: 'Miss Landmine: everyone has the *right* to be beautiful.'

In contrast, Aimee Mullins, who is a double amputee, has branded her body through various prosthetic devices. She has an infamous TED Talk entitled *My 12 pairs of legs*, where she discusses her various experiences with prosthetics—from the Flex-Foot Cheetah used for competing in the Paralympics, to legs that simulate 'natural' feminine appearance complete with (not-so-natural) high heels, to legs that are purely artistic. Interestingly, none of her legs challenge gender or racial corporeal borders. Both women in the Miss Landmine competition and Aimee Mullins challenge the associations of able-ism with beauty while remaining connected to conventional methods of performing or representing corporeal 'beauty' through cosmetics, computer Photoshop technologies, clothing, and various ways of posing for photographs.

Conventional notions of feminine performativity (make-up, clothes, hairstyles, and poses and posture) are maintained or heightened through the pageant and Mullins' commercials for *Cover Girl*. Therefore, despite the challenges to disabled stereotypes and expectations of the body, these examples rely on and reinforce existing and often hyper-feminine beauty regimes.

Activity

1. Compare and contrast Miss Landmine (www.miss-landmine.org) with Aimee Mullins (www.aimeemullins.com).
2. Consider the different corporeal geographies represented and discussed through the images and text descriptions provided on these websites.

Dressing the body

In addition to the various corporeal markers already discussed in this chapter, dress remains another method of expressing identity. It has been, and continues to be, a way for people to associate themselves with part of a larger group, identity, ideology, or occupation. For instance, uniforms are used in different professions as a way for individuals to be easily identified with a particular company or type of work. Sports teams use dress and colors to mark the bodies of players and fans. Clothes can be a method of advertising certain products or fashion labels through effective use of logos and corporate identity markers (Klein 1999). Individuals use dress to signal forms of affinity and belonging with a group, while for some it is also a method of expressing one's individuality.

Religious practice or affiliation can also be expressed through dress. For instance, the Amish (a conservative Christian religious sect in the US) avoid the use of technology as part of their beliefs, and men and women wear specific clothing associated with their gender and age. This form of dress marks them as part of the Amish community, and provides a spatial border between their bodies and those of the 'English' (or non-Amish) living in the same spaces. Amish women cover their heads as part of their religious practice and men grow beards once they marry as a corporeal signifier of that commitment. (For more on religion and dress, see Arthur 1999.)

Some countries have mandated dress codes to initiate wide-sweeping changes to social and cultural norms, or as a method of reinforcing their power and authority over citizens' bodies. For example, modernization efforts in post-Ottoman Turkey, under the tutelage of Mustafa Kemal Atatürk (1923–38), included banning the fez (a popular hat worn by men) and encouraging women to abandon the veil, as both were symbols of the Ottoman past. In contemporary Turkey, veiling has been a contentious site of conflict, resulting, for instance, in the ban of the veil/headscarf in public institutions (such as public schools and universities) because it is a marker of Muslim identity. At the same time, other aspects of Turkish society include the promotion of veiling fashion in an effort to capitalize on Muslim women's sartorial practices. Banu Gökarıksel and Anna Secor (2008) argue that veiling fashion reveals a 'sliding gap between the signifier (the veil) and its desired signification (Islamic womanhood)' (15). Veiling practices within the Islamic context are generally associated with corporeal piety. The veil (and similar corporeal coverings) communicate that the body is a private space, and not to be seen by non-family members or strangers in public spaces.

Veiling fashion is an example of a dialectical corporeal geography. Dialectical arguments juxtapose contradictory ideas. In this case, veiling represents modesty, piety, and privacy, while fashion is an expression of style, consumerism, and public presentation. By bringing these incongruent meanings together (in the broader contexts of Islam, modernity, and capitalism), veiling fashion remains a controversial and contested site of multiple, competing, and (at times) complementary identities. The veil has become a political issue in other countries, such as France, where veiling (along with other corporeal representations of religion) has been banned in public institutions (such as schools). Other countries require women to wear certain forms of dress in an effort to protect them from public harassment or to support Islamic praxis (e.g., Saudi Arabia). Therefore, as several feminist scholars have argued, it is not the veil in and of itself that is problematic, but how governments (or other powerful groups) use dress as a form of sociopolitical and corporeal control. Dress operates as a way of marking corporeal identities, which is malleable and often manipulated for various social, cultural, economic, and political purposes.

Prisons and migration

Other modes of dress used to control bodies can be found in prisons, where inmates are required to wear clothing of the same color, labeled with a number—rather than a name—to identify the prisoner. Prisons are spaces that express the **bio-power** of governments/jailers. French philosopher Michel Foucault used the term 'bio-power' to explain institutionalized practices that regulate and control bodies. There are many ways in which governments control our bodies, such as through public health programs, the regulation of heredity, compulsory education, and law enforcement. Prisons are spaces that represent an acute form of bio-power, where those in power control every aspect of the prisoner's corporeal needs (food, shelter, hygiene, and clothing), schedule, and movement. The regulatory systems within prisons use various mechanisms of corporeal control, discipline, and punishment to contain inmates.

Geographic research on prisons challenges conventional notions that these spaces create social order by detaining and controlling deviant bodies. Geographers such as Ruth Wilson Gilmore (2007) and Jenna Loyd et al. (2012) explicate the ways in which prisons have become structures of oppression against poor, predominantly non-white, and disenfranchised bodies. Gilmore's research on prisons interrogates the growth of the incarcerated population in California specifically, and within the US more broadly. Her research has shown how gross inequalities based on the intersections of race, class, and gender have expedited the growth of prisons through shifting laws and inadequate legal representation for the accused:

> To fight against prison expansion, we would, by joining forces, also have to fight on behalf of clean water and adequate schools, and against pesticide drift, toxic incinerators, all of that stuff. This raised anti-prison organizing in that region [California] to a true abolitionist agenda, which is fighting for the right of people who work in the Central Valley to have good health and to secure working conditions and not be subject to toxicity, in spite of the fact that so many of the workers in the valley are not documented workers.
>
> (Gilmore interview in Loyd et al. 2012, 52)

In this interview, Gilmore links prison growth with declining job opportunities and increasing class divisions between the rich and the poor. She conceptualizes prisoners' bodies as a 'racial' category by the way in which they are marked and treated through the bureaucratic systems that incarcerate them. Rashad Shabazz (2012) further investigates the racialized dimensions of incarceration by analyzing the ways in which diseases such as HIV/AIDS have spread among Black inmates. He identifies the increase of Black male bodies in prison as a form of 'forced migration' due to extensive disinvestment in urban communities, particularly in African American majority neighborhoods. His research shows how carceral punishment has become normalized within US society and how it does not 'address the ways in which Black bodies are marked as criminal nor the profound economic realities that land many Blacks in prison' (298). Gendered analyses examine the rising number of women prisoners as part of a larger process associated with the criminalization of migration and the continued colonization of women's bodies (Walia and Tagore 2012). Incarceration, then, is a sociopolitical and geographic process shaped by racialized, classed, and gendered axes of power and exclusion. And these systems of bodily appropriation and control link up with other feminist geographic critiques of how global processes mark certain bodies in particular ways.

Indeed, labor and labor migration is another way in which bodies are marked, controlled, and monitored as they traverse different spaces through the processes of seeking and attaining employment. Corporeal geographies are an integral part of migration research. For example, Geraldine Pratt's (2004) research on Filipino domestic workers in Canada identifies a lack of rights and a 'loss of bodily integrity' experienced by these workers (96). Because domestic workers often live 'in fear of being accused of theft and feel relatively defenseless around this charge,' they would make sure they did not touch valuable objects in their employers' homes in order to 'leave no trace' of their presence (97). Their migrant bodies, marked by gender, race, and class, were already seen as suspect (or guilty), which caused them to fear being accused of stealing if an item were to be misplaced. Therefore, 'leaving no trace' of their bodies on objects was done to ensure against a potential accusation. Pratt discusses this as a 'type of leakiness across the borders of the body that can be read as both symptom and source of her [the domestic worker's] insecurity about rights' (97). Pratt argues that 'rights confer and protect bodily integrity and the ideal body, seen as "deserving" of rights, is perceived as a "self-contained autonomous body"' (97–8). Therefore, migrant laborers, when perceived as in need or dependent on employers or governments for work and assistance, have often been viewed as 'lacking' the characteristic of the ideal body.

These corporeal representations have the power to separate people from their ability to access rights. Risa Whitson's (2011) research on *cartoneros* (garbage scavengers) in Argentina explicates the ways in which the bodies that work with waste are defined, in relationship to how waste is devalued and separated from society. Because waste is separated from daily living, individuals who handle waste are often either socially or spatially marginalized. Other conceptualizations of corporeal geographies examine the treatment of bodies as laborers within factories: for example, the ways in which women's bodies in *maquiladoras* (manufacturing companies in Mexico) have become worn out (seen as 'waste') through demanding and injurious labor conditions (Wright 2006). The continual turnover of employees in these spaces helps to prevent labor organizing and the ability of workers to demand higher wages and better work conditions. The wasting of female factory workers has become valuable to companies who rely on high turnover as a key feature of their business models (Wright 2006). Refugee and immigrant bodies experience multiple and, at times, compounded experiences of marginalization in different locations.

The treatment and experience of bodies fleeing conflict and seeking asylum includes various layers and experiences of displacement and corporeal uncertainty associated with border management and government bureaucracy (Mountz 2011). Detained bodies through prisons, and those fleeing conflict or seeking work through migration, are continually managed through various forms of governance. Governance processes, procedures, and structures seek to maintain and control bodies in ways that build upon existing racial, gender, and class hierarchies of inclusion and exclusion. The movement of bodies through space, or detained in a specific place through the machinations of power and authority, must continue to be examined from the perspective of individual bodies in order to challenge processes that lead to abuse and the dispossession of rights, and social, political, and economic access to spaces.

Conclusion

This chapter has outlined the various ways in which corporeal markers such as gender, race, class, sexuality, and ability are layered with social, political, and economic meaning. Corporeal geographies examine the ways in which bodies are spatially constituted and organized. The exclusion from, or inclusion of, bodies in places (or their access to rights and opportunities) have been (and continue to be) partly demarcated by one's corporeality. Gender is performed on the body by repetitive acts of doing, rather than being based solely on one's biology. Similarly, identifying and categorizing bodies based on corporeal markers is a dangerous form of social control. These identifications and categorizations have been (and continue to be) used to justify enacting violence, killing, containment, and other forms of oppression on certain bodies. Making social, political, or economic judgments based on appearance limits our ability to engage with others by denying the multiple, complex, and complicated aspects of identity. Use the activity in Box 2.4 as a way to challenge yourself to think critically about corporeal markers and geographies.

Box 2.4 The corporeal marker assignment

(from Fluri and Trauger 2011, 563)

Activity

Corporeal marker assessment requirements

1. As a group, choose and create a visible corporeal marker.
2. Bring in a prototype of your corporeal marker to class. Discuss each group's markers as a class and approve them collectively.
3. Wear this corporeal marker in at least ten different public places for significant time periods, both on and off campus, over the course of two weeks.
4. Wear this corporeal marker both individually and with your group. For example, wear the corporeal marker on the way to meet your group, or on the way home from being with your 'marked' group members. Observe any differences in how you feel and the reactions you receive from others based on your presence in public spaces as a corporeally marked individual, and in comparison to when you are with your corporeally marked group. You will be graded based on your individual

journals, your reflection papers, the design and implementation of this activity, and your responses to the following questions:

- Explain the ways in which your corporeal marker experience created a spatial boundary.
- What were the reactions of individuals (who were not marked) to your presence in public spaces, on and off campus?
- Compare and contrast your experiences when you were marked and alone, versus marked with your fellow group members? Imagine your corporeal marker as one you either cannot or choose not to change, and one that is considered socially, politically, or economically abhorrent in public space. How would you negotiate through public space?
- Consider the implication of this corporeal marking if it were used to identify you as a member of a specific group and to subsequently monitor and/or control your mobility in public space.
- Discuss this assignment in relation to the course material.

Recommended reading

Browne, Kath. 2004. 'Genderism and the Bathroom Problem: (Re)materialising Sexed Sites, (Re)creating Sexed Bodies.' *Gender, Place & Culture* 11 (3): 331–46. doi:10.1080/0966369042000258668

Chouinard, Vera. 1999. 'Body Politics: Disabled Women's Activism in Canada and Beyond.' In *Mind and Body Spaces: Geographies of Illness, Impairment and Disability*, edited by Ruth Butler and Hester Parr, 269–94. New York: Routledge.

Faria, Caroline. 2014. 'Styling the Nation: Fear and Desire in the South Sudanese Beauty Trade.' *Transactions of the Institute of British Geographers* 39 (2): 318–30. doi:10.1111/tran.12027.

Gilmore, Ruth Wilson. 2007. *Golden Gulag: Prisons, Surplus, Crisis, and Opposition in Globalizing California*. Berkeley, CA: University of California Press.

Gleeson, Brendan J. 1999. *Geographies of Disability*. New York: Routledge.

Gökarıksel, Banu, and Anna Secor. 2008. 'New Transnational Geographies of Islamism, Capitalism and Subjectivity: The Veiling-Fashion Industry in Turkey.' *Area* 41 (1): 6–18. doi:10.1111/j.1475-4762.2008.00849.x.

Hansen, Nancy, and Chris Philo. 2007. 'The Normality of Doing Things Differently: Bodies, Spaces and Disability Geography.' *Tijdschrift Voor Economische En Sociale Geografie* 98 (4): 493–506. doi:10.1111/j.1467-9663.2007.00417.x.

Holliday, Ruth, David Bell, Meredith Jones, Kate Hardy, Emily Hunter, Elspeth Probyn, and Jacqueline Sanchez Taylor. 2015. 'Beautiful Face, Beautiful Place: Relational Geographies and Gender in Cosmetic Surgery Tourism Websites.' *Gender, Place & Culture* 22 (1): 90–106. doi:10.1080/0966369X.2013.832655.

Johnston, Lynda. 1996. 'Flexing Femininity: Female Body-Builders Refiguring "the Body."' *Gender, Place & Culture* 3 (3): 327–40. doi:10.1080/09663699625595.

Longhurst, Robyn. 2008. *Maternities: Gender, Bodies and Space*. New York: Routledge.

Loyd, Jenna, Matt Mitchelson, and Andrew Burridge, eds. 2012. *Beyond Walls and Cages: Prisons, Borders, and Global Crisis*. Athens, GA: University of Georgia Press.

McCormack, Derek. 1999. 'Body Shopping: Reconfiguring Geographies of Fitness.' *Gender, Place & Culture* 6 (2): 155–77. doi:10.1080/09663699925088.

McKittrick, Katherine, and Linda Peake. 2005. 'What Difference Does Difference Make to Geography?' In *Questioning Geography: Fundamental Debates*, edited by Noel Castree, Alisdair Rogers, and Douglas Sherman, 39–54. Oxford: Blackwell.

Mountz, Alison. 2011. 'Where Asylum-Seekers Wait: Feminist Counter-Topographies of Sites between States.' *Gender, Place & Culture* 18 (3): 381–99. doi:10.1080/0966369X.2011.566370.

Nagar, Richa. 2014. *Muddying the Waters: Coauthoring Feminism across Scholarship and Activism*. Urbana and Springfield, IL: University of Illinois Press.

Nelson, Lise. 1999. 'Bodies (and Spaces) Do Matter: The Limits of Performativity.' *Gender, Place & Culture* 6 (4): 331–53. doi:10.1080/09663699924926.

Oza, Rupal. 2006. *The Making of Neoliberal India: Nationalism, Gender, and the Paradoxes of Globalization*. New York: Routledge.

Pratt, Geraldine. 2004. *Working Feminism*. Philadelphia, PA: Temple University Press.

Valentine, Gill. 2014. *Social Geographies: Space and Society*. New York: Prentice Hall.

Chapter 3

SPACES OF CULTURE AND IDENTITY PRODUCTION
HOME, CONSUMPTION, AND THE MEDIA

Risa Whitson

Chapter 3

Whitson

Introduction

As discussed in Chapter 2, the body is an important site for the expression of individual identity and group affiliation. In ways that are both within and outside of our control, such as dress, skin color, and mannerisms, our bodies serve as geographic spaces through which we define ourselves and others. And yet it is not only through our bodies that we organize and reproduce social meanings and categories. Rather, the creation of cultural meaning occurs across a multitude of spaces and places. While we tend to think initially of large-scale spaces as they connect to culture (such as national or regional identity), smaller-scale spaces of the everyday are equally important in constituting culture and constructing gendered identity. Places such as houses, malls, bars, ski resorts, and online forums are all crucial in the construction of cultural meaning and the production of gendered, sexed, and racialized identities.

This chapter explores how geographers and feminists consider questions of culture in a variety of spaces. Central to our understandings of the intersections of culture with place, space, and gender are issues of identity, representation, and materiality. In this chapter, we look at these issues in three specific sites: the home, sites of consumption (including tourism and leisure), and the media. While these sites may initially seem unrelated, in each of these three cultural sites, feminist geographers have explored the production of gendered meaning and its connection to power relations. In this chapter, we map these explorations by considering how, in each of these sites, gendered meanings and identities are produced.

This chapter opens with a discussion of the meaning of 'culture,' how it has been used by feminist geographers, and how post-colonial feminists have contributed to our understanding of the concept. Following this, we explore the cultural geographies of the home, focusing particularly on how gendered and sexual identities are rooted in this everyday space. We then examine three sites of consumption—shopping areas, leisure spaces, and tourist sites—to illustrate the ways in which people negotiate and reconfigure gendered meanings in these sites. We finish the chapter by considering how media spaces, including magazines, music, and social media, are also important to the production of gendered cultural meaning and individual identities. In each of these sections, this chapter highlights how these sites are used not only to reproduce gendered and sexualized norms, but to challenge and re-write them as well.

Spaces of culture and identity production

Chapter 3

Culture: identity, gender, and place

Culture has been understood by geographers to consist of 'maps of meaning through which the world is made intelligible' (Jackson 1989, 2). These maps of meaning, which vary over time and space, organize and place value on individuals, objects, and activities. While culture plays a powerful role in giving meaning to people and social practices, culture is not something that exists separately from societies and makes individuals behave in certain ways or believe certain things. Rather, culture should be understood as a **socially constructed idea** that people constantly create through everyday actions, that organizes social practices and makes them appear to be natural or normal. In the same way that race has no empirical (biological) reality but nonetheless confers organization, stratification, and meaning on society, culture also functions as a powerful concept that is used in ways that maintain particular social relations of power. Culture is thus a socially constructed set of practices and representations that gives meaning to people, places, activities, and objects.

In their critique and discussion of cultural geography, feminist geographers Jane Jacobs and Catherine Nash (2003) assert that **cultural geography** is a field focused on 'questions of identity, social construction, representation, positionality, and difference' and the relationship of these to space and place (269). Moreover, because of the centrality of gender to each of these concepts, they argue that cultural geography connects intimately to a feminist framework for understanding society. One of the central functions of cultural processes is categorization and differentiation, and gender is one of the primary ways in which our society produces and naturalizes difference. Thus, gender must be understood as a foundational element of how culture is understood and enacted. For example, as discussed in Chapter 2, in many post-colonial societies, a concern with maintaining women's traditional gender roles and dress can be seen—even in contexts in which traditional clothing or occupations are not expected for men. In these situations, gender norms, expressed through the tangible culture of clothing and the cultural organization of work, serve as visible markers of 'difference' between the 'traditional' and the 'modern.'

Post-colonial feminists, discussed briefly in Chapter 1, have contributed to geographers' understanding of culture through their assertion that the concept of culture has been used by those in positions of power to maintain social hierarchies of gender and place. Authors such as Uma Narayan (1997), Chandra Mohanty (2003b), Lila Abu-Lughod (1991, 2013), and Sara Ahmed (2004, 2006, 2010) argue against an understanding of culture as something that is coherent, homogenous, and static, suggesting instead that cultures are constantly changing and being defined through political struggle. The definition of what is understood as one group's 'culture' and 'tradition' functions to create and fix differences between people. As such, the concept of culture is a principle tool for creating difference, in that it helps us define ourselves in relation to other groups who are seen as 'outside' of our culture.

One example of the connection among culture, power, and identity is how viewing women in post-colonial contexts as 'victims of culture' problematically reinforces the superiority of the West by perpetuating understandings of post-colonial cultures as timeless, unchanging, and ahistorical. Narayan (1997) presents an example that challenges this view in her discussion of *sati* in India. *Sati* is a practice in which widows commit suicide by burning themselves (or being burned) on their husband's funeral pyres. While this practice was never widespread, and is now obsolete, it was nonetheless at the center of discussions around Indian culture and tradition during the last century-and-a-half. Narayan argues that *sati* gained its status as a part of Indian culture only through the investigation of British colonizers into the practice in an effort to determine if it should be outlawed. As she states, '*sati's* "status as tradition" seems to have arisen in the context of colonial worries about whether the practice could be outlawed without eliciting extensive protests from segments of the Indian population' (61). The British investigation into *sati*, and its subsequent outlawing in colonial India, had the effect of elevating the status of *sati* from a practice that was limited to a specific caste group in one small region of India, to one perceived as a tradition that was central to Indian culture. At the same time, it helped to legitimize British rule in India as a civilizing force through reinforcing the perception of the cultural otherness of India and positioning Indian women as victims who needed to be saved from their own culture. In tracing this history, Narayan highlights the socially constructed nature of culture, its role in struggles over social power, and the way in which culture becomes fixed to our understandings of people and place. This example also illustrates the importance of considering the political and economic effects of how we define cultures, or what Narayan refers to as 'the politics of tradition formation' (61). This is particularly important for understanding women's life possibilities in different geographical contexts and the possibilities of changing these for the better.

As this discussion suggests, one of the most critical functions of cultural processes is the production of difference and categorization. The production of difference is often rooted in a distinction between 'the self' and 'the

Spaces of culture and identity production

other' and, as such, is central to understanding the identities of individuals and social groups. As Box 3.1 indicates, this process of differentiation is constant, as we define concepts such as 'American culture' through our everyday experiences and in our everyday spaces of work and play. These questions of identity and difference, and their connection to place and space, have been key in the research of feminist geographers; in particular, feminist geographers have addressed the ways in which particular spaces and places, and activities and behaviors within these spaces, function to anchor cultural meanings and subsequently confer these meanings on individuals and groups.

Box 3.1 The politics of culture

Figure 3.1 Disneyland's Main Street, USA. Reproduced courtesy of Loren Javier.

Activity

Disneyland's Main Street, USA: 'It's as American as apple pie!' Or is it? Examine this iconic American landscape that Disney has created to consider how 'American culture' is defined. Consider the symbols, traditions, and celebrations that make up what we consider to be 'American culture.' Who do these symbols represent? Who is included? Whose experience of 'America' is not included? Consider how 'American culture' is the result of political negotiation that constructs some people as insiders and some as outsiders.

51

There are a number of important sites in which this production of difference occurs. In its role as a boundary that separates 'us' from 'them,' the home (both physical houses and homeland) becomes a central place where cultures are defined and organized. Cultural processes are also closely linked to political (as the example of *sati* indicates) and economic processes. In fact, in contemporary societies, practices of consumption are increasingly used to express identity and cultural difference. Finally, because of the importance of representation to identity and differentiation, the media plays a critical role in shaping, perpetuating, and challenging cultural norms. In this chapter, we focus on these three sites that produce and maintain gendered meaning and identity: the home, sites of consumption, and the media.

The home, gendered meaning, and identity

The cultural norms that regulate gender are strongly related to our understandings of the meaning of 'the home.' For this reason, the home is a key site in the production of gendered difference, meaning, and identity. The **home**, like other places, involves an 'idealized meaning or social imaginary' (for example, the home is love, family, belonging), a 'physical or material reality' (the building of the home, the objects within it, the arrangement of rooms and furniture), and a unique set of 'social relations' (for example, the home is often associated with family, but also with other relations such as friends or roommates). The home is also a 'process,' such that the word 'homemaking' means not simply 'turning a house into a home' (Morrison 2013, 413), but also involves the construction of individual subjectivity as well as a movement toward stability, social inclusion, and belonging. As Carey-Ann Morrison (2013) describes, 'homemaking embodies, reflects and supports subjectivities (individual and collective) through everyday practices such as domestic routines, through social and intimate relationships, and through the accumulation and arrangement of meaningful objects' (413). In this way, the physical home interacts with the social imaginary of the home through people's lived experience to help constitute and support their identities, or sense of themselves (Blunt and Dowling 2006).

These different elements of the home (imaginary, material, and social) all converge in the process of homemaking to produce a site that is critical to the construction and maintenance of gendered meaning. One important component of this is the separation of public and domestic space, which is rooted in an ideology of separate public and private spheres. The advent of urbanization and industrial capitalism in the 1500s to 1700s, paired with the subsequent

Spaces of culture and identity production

normalization of male waged labor outside of the household, transformed the home from a space of work and living for all members of a family, to a space associated primarily with women, children, and non-work activities (Domosh and Seager 2001). This **ideology of separate spheres** remains present today, both in the ideology of the home and in its material expression. Dolores Hayden (1980), for instance, describes the way in which the built environment reflects the ideological separation of spheres in urban design. She argues that land use patterns and zoning laws that separate residential areas from industrial and commercial centers are a way of materializing this separation of work/male space from home/female space on a large scale. The implications of this separation of spheres on urban design and experiences of urban space are further discussed in Chapter 4, while its effect on work and households is considered in detail in Chapter 5.

The ideology of separate public and private spheres not only affects the spatial organization of homes, but also continues to influence our ideals of the home and our expectations of what social relations will be present in the home. These aspects of place in turn shape not only *who* we associate with the home, but also what *activities* are normalized in that space. Advertisements and popular culture serve as constant reminders of what people and activities are normalized within the home. While the division of labor and makeup of the household have changed over time, contemporary advertisements continue to express gendered ideals similar to those in advertisements from previous decades (as discussed in Box 3.2). We continue to see images of happy (female) homemakers cleaning, cooking, and caring for their families in the home, and happy (heterosexual) families playing football and board games, and watching movies together. In the popular imagination, therefore, the home remains a feminized, heterosexual space of non-work. These cultural norms, which feminize the home, continue to have a strong hold, even in contexts in which women commonly leave the home to engage in waged work, and men routinely engage in domestic work within the home.

In the 1960s and 1970s, the feminization of the home and its status as a place of leisure, as opposed to work, were both highlighted and problematized by authors and activists such as Betty Friedan, the founder of the National Organization for Women (NOW), and Pat Mainardi of the radical feminist group Redstockings. Friedan's enormously influential book *The Feminine Mystique*, published in 1963, marked the beginning of the second wave of feminism in the US. Through her discussion of the plight of the American housewife, Friedan recast the idealized suburban home as a space not of leisure

> **Box 3.2 Gendered images of the home**
>
> **Activity**
>
> The Gender Ads Project, created by Dr. Scott A. Lucas, is a website that catalogues the way in which gender and sexuality appear in print media advertisements. Visit the page (www.genderads.com) and click on the tab 'Roles' on the bottom of the page; scroll through the ads collected for this page. Examine the ways in which the images connect men's and women's roles differently to the home. Many of the images included on this website are historical. Consider other advertisements that you have seen recently on television, on the internet, or in magazines for appliances, cleaning agents, and other products used within the home. What do these images say about who is responsible for making the home into a space of comfort? Do the material aspects of the home that are highlighted in these advertisements (such as garbage disposal, small kitchen appliances, and others) contribute to the gendering of home spaces? Do you feel that these advertisements reflect *your* experience of the home? Why have these types of media portrayals of the home remained so similar over the years, even when women and men are increasingly sharing household labor?

and domestic happiness, but of confinement, boredom, tedium, and depression for American women. While Friedan focused primarily on white, upper-class, highly educated women, and as such was critiqued by feminists of color as being racist and classist (hooks 2000), her book brought into question popular understandings of the home that were a critical part of post-war society in the US. Similarly, Pat Mainardi's (2012 [1970]) essay 'The Politics of Housework' redefined domestic labor as work and highlighted the political nature of this trivialized form of labor. As both of these publications suggest, the production of gendered difference and the meanings ascribed to gendered identities are anchored in people's experiences of the home, and the activities that occur in that space. See Box 3.3 for more about gendered spaces in the home.

In addition to the feminization of the home through the public/private binary, there are a number of other ways in which the home plays an important role in the production of identity. The ideal of the home rests on a heterosexual norm; that is, homes are heteronormative spaces that are closely associated with the nuclear family. In spite of the fact that the two-parent, heterosexual, nuclear family is not the norm in many places, including the US

Spaces of culture and identity production

(in the sense that it does not actually describe the average US household), it continues to be normative in our understanding of what 'the home' means. For queer youth, the effects of this conception of the normative home can range from a lack of freedom to perform their identities (i.e., to be themselves) at home, to estrangement, fear, and homelessness. This is strikingly clear in the statistics regarding homeless youth: approximately 40 percent of the youths served by homeless shelters in the US identify as LGBTQ, and almost seven in ten of these youths report being homeless because they were forced out of, or ran away from, their homes as a result of their gender identity or sexuality, or because they experienced violence at home (Durso and Gates 2012). While these types of statistics provide a baseline for understanding the ways in which heteronormative understandings of the home affect the identities and experience of queer youth, feminist geographers such as Andrew Gorman-Murray (2008b) explore how these norms are experienced and may be challenged by the families who accept and affirm their LGBTQ children. Additionally, Lorraine Young (2003) delves deeper into the question of the production of cultural meaning through home spaces, by exploring how homeless youth resist normative meanings of the home and actively create positive identities in the context of homelessness.

Box 3.3 Gendered spaces in the home

Activity

'Don't sit in Dad's chair!' 'Get out of my kitchen!' These are phrases that many of us grew up hearing and that reflect the gendered organization of home spaces. Think about the house where you grew up: in what ways was it organized as a gendered space? You may want to consider the following questions:

1. Who was responsible for the work in the house and in the yard?
2. Were all of the rooms in the house used by all family members (the kitchen, the den, the TV room)?
3. Who decorated each room? Whose personality and interests did the decorations reflect?
4. Were there particular spaces reserved for some members of the house?

Now think about the spaces you live in now, such as your dorm, apartment, or house. Have you replicated or challenged the gendering of spaces that you grew up with? How does the organization of space in your home reflect who you are?

While queer youth often experience the home as a place of exclusion, for gays and lesbians with the resources to make a home their own, this space presents a possibility for the expression of alternative sexual identities that may not be available in other venues. For older lesbians and gay men in particular, the home becomes an important site where queer identities are articulated and valued, and, as a result, where normative understandings of domesticity are challenged and reshaped. For example, Brent Pilkey (2014) uses interviews with older gay men to explore ways in which objects in their home may serve to both reinforce and challenge heterosexual norms connected to domestic space. He identifies material objects such as 'diva' CDs, rainbow-colored decorations, and mementos from gay couples' family vacations, as well as same-sex family photos, including wedding photos, as ways in which the materiality of the home is used to express identity, as well as reinforce and challenge normative understandings of domesticity.

In addition to being *hetero*normative, the popular imaginary of the home is normative in other dimensions, as it expresses a racialized and classed ideal. That is, the idealized home is imaginary, based in a middle-class, white, privileged experience. This means that many people do not have the opportunity to create an experience of the home (physical or social) that approximates the ideal. For example, consider the experiences of the meaning of the home and the process of homemaking among refugees, those dwelling in informal settlements, and the homeless. These cases illustrate how the inconsistencies between the ideal of the home and the lived experience intimately affect the development of individual and group subjectivities, as these differences serve to systematically marginalize certain populations and ways of life.

Yet even for those who live in physical homes that approximate the ideal, the actual experience of the home can be far from the expectation (Brickell 2012). For example, women and others suffering from violence in the home may experience it not as a place of love, belonging, and comfort, but as a site of loneliness and insecurity. Similarly, for people with disabilities, 'people's domestic experiences are, potentially, at odds with the (ideal) conceptions of the home as a haven, or a place of privacy, security, independence, and control' (Imrie 2004, 746). Instead, disabled people may experience the home as a site of isolation, entrapment, or insecurity, as it may reinforce their own lack of control, privacy, and autonomy.

As discussed further in Chapter 6, feminist geographers have challenged an apolitical understanding of the home which casts the 'public' as the sole space of political action. Meghan Cope (2004) describes domestic spaces as

Spaces of culture and identity production

critical for women's political activism, as women and other groups without access to public spaces are able to use domestic space precisely because it is coded as 'women's space.' In these domestic spaces, women and others are therefore viewed as apolitical, and consequently able to engage in organization and resistance that may not have been possible in public spaces. bell hooks (1990) clearly expresses this in her description of the ways in which the home served as a safe space that fostered humanization, affirmation, and dignity in the context of a racist society and public sphere. She argues that in spite of 'sexist thinking about the nature of domesticity,' which results in the home being 'conceptualized as politically neutral space' (45, 47), by providing a space where African Americans can be loved and respected for who they are in the context of oppression and domination, the home functions as a site of subversion, resistance, and liberation struggle.

Finally, as feminist geographers have recently argued, masculinities and male identities must also be understood as closely connected to this important space. The term **masculinities** refers to the position that men hold within a gender system, along with the expectations, performances, and patterns associated with this position among both men and women (Connell 2005; Berg and Longhurst 2003). Because of the socially constructed nature of the gender system, masculinities are always multiple and contingent, meaning that they vary according to time and place, and are expressed and lived differently as a result of different intersectional identities. As such, rather than the home being simply a space connected to women's identities, it is the case for men as well that 'as one "makes home", one accumulates a sense of self' (Gorman-Murray 2008a, 369).

Gorman-Murray (2008a) argues that there are distinct types of domestic masculinities—that is, masculinities associated with, and constructed through, the space of the home—including those represented by men in heterosexual unions, bachelors, heterosexual men sharing home spaces with other men, and gay men. As described above, geographers have focused on the domestic masculinities of gay men through explorations of how the process of home-making allows for the expression of alternative identities and the revaluation of marginalized identities. At the same time, a number of geographers have also considered the ways in which domestic masculinities are articulated for men who identify as straight. Being an involved dad, a handyman who is good at fixing things up, or a single man whose 'bachelor pad' reflects his interests are all distinct ways in which the space of the home contributes to masculine identity (Cox 2015). While these identities at times reinforce the conventional understanding of men as 'breadwinners' in relation to the home, increasingly these domestic masculinities also allow men to 'negotiate alternative masculinities, where they could be expressive, emotive and engage in domestic labor and child care' (Gorman-Murray 2008a, 369). Thus, although the home is generally labeled as a feminized space, it works to inform identities and create social meaning across the gender system.

Sites of consumption: shopping, leisure, and tourism

While the home has been explored as an important site of meaning and identity production in feminist geography since the 1980s, within the last two decades, **sites of consumption** have also been analyzed as increasingly important to cultural geographies of gender. 'Sites of consumption' can refer to a variety of places and activities, as the process of consumption involves a myriad of goods and services. Through the process of consumption—which might range from buying a mocha latte at Starbucks, to wearing an iWatch, to picking something up at the thrift store, to wearing your favorite team's jersey, to traveling on vacation, to going out dancing—people construct and perform their social status, group participation, and individual identities.

While consumption is usually connected to spending money and acquiring products and services, Wanning Sun (2008) reminds us that it is much more than that. In her investigation of the consumption practices of female Chinese rural migrants living in Beijing, she describes the way in which sites like the supermarket can become 'places of refuge where new arrivals seek to cope with their fear of the city and sense of isolation' (483). While the supermarket is a mundane space of consumption, it also becomes a place that enables the migrants to battle 'loneliness and fear' (483). Similarly, we can think of the way in which teenagers 'hang out' at the mall to engage in social group activities (even if they have little money to spend) as an example where a site of consumption serves as a significant space of entertainment and leisure. As is clear from these examples, sites of consumption are important not only for the wealthy, but also for those with few financial resources. In this way, spaces of consumption offer much more than an opportunity to engage in an economic transaction; they constitute 'a medium through which to read identities based on social differentiation and value' (Hofman 2013, 1010).

Shopping

Because of the connection of consumption to identity formation, feminist geographers are particularly interested in how all types of consumption connect to the maintenance, performance, and negotiation of gender and sexual norms and identities. Experiences of shopping provide one lens through which to explore these questions (Gregson et al. 2002). Indeed, shopping has been 'intimately connected with the social and symbolic constructions of gender and spatial relations' from the mid-1800s onward, when, as masculinity came to be more strongly connected to work, the concepts of leisure and consumption became feminized (Rabbiosi 2014, 212). Not only are women's gendered domestic responsibilities often connected to shopping activities, but markets and shopping malls are often designed to appeal to, and promote, particular versions of femininity. For example, Chiara Rabbiosi's (2014) study of high-end outlet malls in Italy concluded that competing gendered narratives of women's sensuality, efficiency, and motherly care are reproduced through the advertisements at these shopping malls, and women's practices in these spaces both conform to and, in some ways, modify these roles.

Figure 3.2 Shopping malls as feminized spaces. Reproduced courtesy of Woody Wood.

As discussed in Chapter 2, clothing and dress are critical ways of marking the body to express individual and group identity. The connection of clothing to identity also occurs through the practice of consumption and, as Rachel Colls (2004, 2006) argues, emotions play an important role in shaping the experience and meaning of such consumption activities. She challenges academic analyses of women's shopping experiences that present women as either victims of, or resistant to, gendered bodily norms. Colls (2006) identifies how 'clothing and the spaces of clothing consumption act as the creative and materializing medium' (530) that enables women to emotionally experience their bodies in a number of ways. For instance, it can allow them to work through particular emotions in order to feel comfortable about their bodies. Colls (2004) describes this process as follows:

> Elisabeth and Jamie reflect on the inconsistency of numerical sizing so that they can be a size smaller than they really are, in order to feel better about their size. Joanna uses a tape measure to avoid trying on clothes that will not fit her when she tries them on at home, as she attempts to cope emotionally with her recent weight gain. Finally, Rosa connects with a 'kindred' body in order to articulate how her feelings about her body and her practices of dressing have changed. Whilst she still recognizes that her self-esteem is 'still through the floor', Rosa uses this particular practice to enable her to feel better about herself.
>
> (593)

As such, consumption not only demands conformity to social expectations, but also provides the possibility of destabilizing these expectations through individual practices and emotional negotiation.

Consumption is not simply a gendered activity, however. What we do when we shop and how we experience different spaces of consumption are also influenced by our race, class, age, and sexual expression. Petra Doan (2010), a transgender woman, highlights this as she describes one of her first attempts to pass as a woman in a shopping mall. Rather than being an uneventful trip to the mall, Doan expresses the anxiety that she felt as she had a sense of 'being watched' and feeling 'stares boring into [her] back.' After this experience, she concludes that 'malls are clearly not places that welcome gender transgressive behavior' (645).

Elizabeth Chin's (2001) ethnography of low-income African American children's consumption practices also underscores the exclusionary nature of spaces of consumption. Nonetheless, she also demonstrates the ways in

Spaces of culture and identity production

which, even for people whose purchasing power is low, the process of shopping is connected intimately to the creation and maintenance of identity, social networks, and kin relationships. For young African American girls, this occurs through activities such as negotiating a store clerk's suspicion, using limited funds to buy gifts for family members, or considering how (thin, white) Barbie fits into their lives. By observing these children's consumption practices, Chin concludes that 'the forms of social inequality that come to bear upon children in different consumption sites are a primary factor in shaping their experience in and understanding of those places; in that process, children also build apprehensions of themselves and the society in which they live' (92).

Doan and Chin's research suggests that sites of consumption are important spaces for social interaction. This is particularly the case in contexts in which malls, markets, and other shopping areas serve as the primary public spaces that are available, or most desirable, to a population. In places as diverse as Nairobi and Bogota, for example, many middle- and upper-class people choose to recreate inside malls due to concerns that parks and other open public spaces are unsafe (particularly for families). Chapter 4 discusses how malls can also serve as safe spaces for women to socialize in contexts such as Tehran, where women's presence and mobility may be restricted in other types of public spaces. It is important to note, however, that these interactions are circumscribed by class and race relations. As privately owned spaces, shopping malls can restrict access and refuse entrance to individuals and groups who are not seen as complying with the norms established for these spaces. Furthermore, these spaces often serve to informally segregate populations by class and race. In Buenos Aires, for example, poorer people will shop at outdoor markets (Figure 3.3), while the middle class frequent shopping malls such as Buenos Aires Abasto. The high-end shopping areas, such as Galerías Pacífico and Alto Palermo in Buenos Aires, are frequented primarily by tourists and wealthy locals (Figure 3.4). As a result of the intersection of race and class privilege with gender experience, for women without race and class privilege, malls do not function as a safe social space. Rather, they may be experienced as a place of exclusion and anxiety.

Leisure

While routine shopping for food and clothing is one important site of consumption, leisure activities constitute a second critical site of consumption. Public leisure spaces, such as sports venues, parades and festivals, and spaces of nightlife

Figure 3.3 An open-air market in Rosario, Santa Fe, Argentina. Reproduced courtesy of Maximiliano Kolus.

Figure 3.4 Galerías Pacífico mall in downtown Buenos Aires. Reproduced courtesy of Martin St-Amant. Wikipedia – CC-BY-SA-3.0.

like clubs, bars, and pubs, are sites of particular importance in the construction and negotiation of gendered and sexualized meaning (Caudwell and Browne 2003; Tan 2013). For example, the regulation of pub spaces historically worked to constitute pubs as masculine spaces, and to functionally prevent women's use of these spaces (Beckingham 2012). In a contemporary context, these public spaces are also ones in which women may be more likely to conform to heterosexual feminine norms than other spaces, such as work or home spaces. For example, while young women in Beijing describe challenging traditional gender norms in private spaces with their partners, in public leisure spaces, such as bars and nightclubs, they continue to reinforce traditional norms of femininity and masculinity (Jin and Whitson 2014). Young women do this by restricting certain behaviors (such as drinking and smoking in public, or attending a bar or dance club alone), while actively working to support their male partners' masculinity. Similarly, Boyd (2010) describes the ways in which mainstream nightclub venues in Vancouver promote hyper-femininity and masculinity, simultaneously reinforcing strongly heteronormative spaces (Figure 3.5). In addition to women's gendered performances being monitored in public leisure spaces, Phil Hubbard (2013) also argues that other social groups (especially youth, the working class, racial and ethnic minorities, and sexual minorities) 'are rendered problematic in night time leisure spaces, with their bodies becoming the focus of surveillance and media scrutiny' (277).

Figure 3.5
Bars are examples of public leisure spaces where gendered and sexualized meaning is produced. Reproduced courtesy of Woody Wood.

As the above examples illustrate, public leisure space is also an important context for the construction and maintenance of masculine identities. Morin et al. (2001) explore this in a historical context by looking at the masculinist, nationalist, and racialized construction of mountaineering in the late 19th and early 20th centuries. They argue that mountaineering was perceived as a sport for which 'macho bravado' came to be valorized through a focus on challenge, adventure, and the need for inner strength and self-discipline (121). As such, they state that 'the story of the first completed ascent of Mount Cook [New Zealand's tallest mountain] reads like a competition of manly, national interests, and as such expresses the earliest association between a masculine mountaineering identity and nationalism' (123). In a more contemporary context, feminist geographer Gordon Waitt (2008) has also traced the importance of spaces of leisure to masculine identity formation, through examinations of surfing, bodyboarding, and related socializing in Australia. Waitt describes the way in which surfing not only provides a space for men to socialize with other men, but also by being in a high-risk activity, surfing also becomes a testing ground, 'enabling acts of heroic masculinity' (84). These masculinities are also explicitly spatialized, as surfers defend particular territories where they are expected to deal 'with physical assaults and verbal insults "like a man" . . . helping to fix the taken-for-granted assumption of the surf as a masculine social space' (86). In other contexts, although women increasingly attend high-revenue sporting events, such as football, soccer, basketball, and baseball games, these events continue to reinscribe dominant forms of masculinity (Figure 3.6).

While spaces of leisure reinscribe dominant norms of gender, sexuality, and race, these venues can also become sites of resistance to gendered, sexualized, and racialized expectations. Tiffany Muller (2007) provides one example of this in her description of lesbian 'kiss-ins,' which seek to make lesbian identities visible in the space of a women's professional basketball team, the New York Liberty. Muller writes:

> Tired of seeing lesbians in the majority in the stands and yet ignored by Liberty management policies, Lesbians for Liberty protested by drawing attention to lesbian attendance at the game, most notably by standing and kissing during breaks in play, and by waving banners stating, 'Liberty: Lesbian fans fill your stands. Liberty for All?'
>
> (197)

Spaces of culture and identity production

Figure 3.6 Fans at a Philadelphia Eagles game. Reproduced courtesy of Woody Wood.

LGBTQ pride parades and festivals have also been identified as particularly important sites of leisure for challenging normative expectations about sexuality (Johnston 2005). Indeed, Kath Browne (2007) refers to these events as 'parties with politics,' recognizing that one of the factors that makes these events so powerful in challenging norms is 'the centrality of the party, fun, emotion, and hedonism in discussions of political effects' (82).

Tourism

Spaces and experiences of a third form of consumption—tourism—also offer insight into how gendered, sexualized, and racialized norms and behaviors are invested with power, particularly in a post-colonial context. Because tourism involves moving outside of local gendered norms, it can provide freedom from gendered expectations for marginalized groups in ways that may promote feelings of empowerment, as well as increase feelings of social exclusion and vulnerability. In this way, while 'girlfriend getaways' may allow women to escape repressive norms, in other contexts, unaccompanied women tourists may feel conspicuous and unsafe.

Sex tourism is a particularly powerful site for considering how constructions of not only gender, but also race and nation, are engaged through consumption processes. One reason for this is that the meanings that travelers associate with particular destinations work in conjunction with gendered and racialized stereotypes of 'native' masculinities and femininities to reproduce imperialist power relations through intimate relationships. As such, while we discuss the connection between tourism and identity in this chapter on culture, it also serves as an important example of the geopolitics of race, gender, class, and sexuality discussed in Chapter 6.

Sex tourism constitutes a 'multi-billion dollar transnational industry' which creates employment for the men, women, and children who are employed as sex workers (Mullings 1999, 58). Sex tourism is also connected to other jobs (brokers, finders, operators) and businesses (sex shops, massage parlors, exotic dancers, websites for sex tourists, etc.) that are connected to the industry. While both men and women participate in this industry as sex workers, male tourists and female sex workers remain the norm. As Mullings (1999) argues, 'like the new Third World assembly industries, the current international division of labor in the sex tourism sector primarily relies on the bodies and labor of women of color to create wealth' (57). This has been connected to the performance of 'imperial masculinity' among sex tourists, functioning through 'the real and fantasized subordination of women,' which constructs and naturalizes 'ideologies of racial, class, ethnic, and sex/gender differences that both registered and reinscribed the sociospatial hierarchies of the global division of labor' (Gregory 2014, 133).

While sex tourism is generally associated with male tourists and local women, feminist geographers have also explored women's sex tourism and gay sex tourism. In their analysis of gay sex tourism in the Caribbean, Gordon Waitt et al. (2008) argue that:

> Drawing on racialized fantasies regarding the sexuality of Caribbean people, travel guides remain filled with suggestions that the only salient characteristics of men and women of colour . . . are their apparent 'animalism', sexual fluidity, pre-capitalist social order and pristine living environment. Characterized by the modernity/tradition split, the Caribbean is portrayed . . . as a paradise free from homophobia and offering sexual freedom, free sex and taboo sexual encounters.
>
> (786)

Similarly, in the matriarchal Mosuo community of China, 'the consumption of sensuous satisfaction is grounded in imaginative discourses which construct

Spaces of culture and identity production

the Third World travel destinations into places of less moral constraint, of men and women with hypersexual attractions, and of different value systems of "shoulds" and "oughts"' (Qian et al. 2012, 109). Sexual encounters with locals in tourist destinations are thus connected to the meanings of a place in such a way that 'the sex/love object is not just the [local person] but also a landscape' that is imbued with historical and political meaning (Jacobs 2009, 44).

Locals in tourist destinations in the Global South also become a part of the active negotiation of these relationships, often to their perceived benefit. In addition to providing an economic livelihood, tourist-oriented sex work can provide an opportunity for individuals to increase their access to financial resources and interpersonal networks, and thus improve their possibility for social mobility. These opportunities may come through close relationships with foreigners that result in long-term personal commitments, increased integration into the English-speaking community, sponsored migration, greater cultural capital, or business opportunities. In this regard, Lisa Law (1997) and Jessica Jacobs (2010) explore very different examples of sex workers—with Law examining female sex workers in the Philippines and Jacobs considering male sex workers in Egypt—but reach very similar conclusions. As Law (1997) argues, through relationships with Western men, women are perceived as 'becoming modern' (122), which provides a social mobility that is distinct from the economic benefit of their work. Similarly, Jacobs (2010) argues that for male sex workers in Egypt entering into relationships with American and European women, one goal is to 'access the mobility of the Western citizen (and the positionality of the modern male subject)' (77). She thus argues that if relationships between female tourists and local male sex workers 'were an escape from modernity for many women, for men they offered an escape *to* modernity' (77). Furthermore, sex tourism is connected not only to the goals and aspirations of individual sex workers, but can also be integral to the economic development and modernization goals of communities and countries, as local communities may also support sex tourism 'in pursuit of imagined progress and modernity' (Qian et al. 2012, 107).

Overall, spaces of consumption highlight the ways in which relationships of power permeate cultural processes in our everyday, lived spaces. Shopping spaces, leisure spaces, and tourist destinations are all spaces in which gendered, racialized, and classed systems structure our experiences and opportunities. These spaces also provide the material through which identities are formed and the 'self' and 'other' are defined.

The media and feminist geographies

The **media** is a third site that is central to the production and circulation of cultural representations. Over the past two decades, feminist geographers have considered a variety of media forms with two general interests. First, they use the media to explore the gendered and sexualized cultural meanings that circulate within society; in this way, the media serves as a lens through which to better understand gendered norms, ideals, and expectations. Most frequently this involves studying newspaper articles, but geographers have also used other forms of media such as television, movies, music, and literature, and more recently **new media** such as websites, chatrooms, social media, and digital gaming. Second, researchers also examine these forms of media as sites that are themselves active in the production, distribution, and negotiation of these cultural norms. From this perspective, the media itself becomes the focus of critical inquiry.

Examples of research focusing on representations of gender across traditional media genres include Carol Morris and Nick Evans' (2001) examination of British farming magazines, Marcia England's (2006) study on horror films, and Jason Dittmer's (2005) analysis of Captain America comics. These studies find that the media serves to reproduce gendered binaries of 'hegemonic masculinity and emphasized femininity,' albeit in 'increasingly subtle and fragmented ways' (Morris and Evans 2001, 375). These expressions of masculinity and femininity are always also intersectional in nature; that is, stories that are told about gender through the media are also stories of class, race, and nationhood. Margath Walker's (2005) research underscores this intersectionality. Drawing on a discourse analysis of articles published in the national Mexican newspaper *La Jornada*, Walker argues that portrayals of women as anonymous, replaceable bodies, as victims of the border city, and as 'dependent appendages' all work together to construct 'women as weak, sexualized objects, and Mexicans as raced, backward "others"' (95). In this way, narratives of masculinity and femininity are produced and circulated through the media in ways that reinforce conventional and problematic understandings of gender, ethnic, and national identity.

In addition to examining traditional media, during the last two decades feminist geographers have been increasingly engaging media use and representations in a new type of space: cyberspace (Rose 2016). According to Nicky Gregson and Gillian Rose (2000), rather than being understood as separate, geography and cyberspace should be understood as intimately connected. To this end, they use the term **cyber/space** to highlight the way in which these

Spaces of culture and identity production

two types of space interact to form a hybrid space, in which the boundaries of both the geographic and the virtual are blurred.

Drawing on this concept, Clare Madge and Henrietta O'Connor (2005) have conducted research into online mothering forums which has found that although cyberspace offers a new degree of anonymity and the ability to connect across distances, it does not replace geographic space. Rather, because we cannot leave our physical lived and embodied experiences behind when we enter cyberspace, virtual interactions continue to be 'grounded in the social, bodily, and cultural experiences of users' (85). Nonetheless, they assert that in online mothering forums, mothers are able to 'try out' new subject positions and identities that may be constrained in the physical world. As such, they argue that cyberspace is a **liminal space**, as it represents a space of between-ness, grounded in the corporeal, but offering creative possibilities.

Because of the creative possibilities that cyberspace offers, some researchers believe that there may be more space for alternative representations of gender and sexuality in new media and social media than in traditional media. Jessica McLean and Sophia Maalsen (2013), for instance, have examined the revitalization of feminist movements on and through new media in Australia. They argue that new media holds a number of opportunities for mobilizing: it allows multiple voices to be expressed and, as such, is more egalitarian; it enhances connectivity; and it facilitates the connection of personal lived experience to political activism through the semi-anonymity it provides. In particular, they argue that new media enables political action to be complemented and supported by personalized politics, as it transcends the boundaries of public and private spaces. In this way, these authors argue that the liminal nature of social media produces a 'quasi-public/private space' that has the possibility of producing powerful interest groups and bringing awareness to critical feminist issues.

Additionally, for those who have access to the internet, venues such as blogs, social media, forums, and alternative news sources provide a potential space for the expression of ideas and the dissemination of information not represented in mainstream media sources. Oreoluwa Somolu (2007), for example, argues that for African women, blogging provides a venue in which they can make content available that addresses their interests and meets their needs. She argues that this venue, in turn, is a powerful tool of social change because of its ability to

> give a voice to the previously unheard, and provide them with the tools to connect with others who share the same concerns, while reaching out to

people who might have been unaware of these issues, and providing them with a platform on which they can map a strategy for raising the quality of women's lives in Africa.

(487)

Similarly, some researchers highlight the ways in which social media is now being used by marginalized groups to connect, organize, and empower themselves through platforms such as Hollaback! (discussed in Chapter 4), Black Lives Matter, and political organizing, such as that which occurred during the Arab Spring. As such, social movements like the World Social Forums, which have been held annually since 2001, rely increasingly on information and communication technologies (ICTs) for network building and coalition organizing. A widely publicized case of new media and social media being used to promote gendered justice can be found in the example of the activism of Manal al-Sharif, a Saudi woman who posted a video of herself driving as part of her activism to legalize women's driving in Saudi Arabia. This video, and the accompanying movement (Women2Drive, which supports women's rights to public mobility in Saudi Arabia, including the right to drive), gained enormous global popularity through social media channels, including YouTube and Facebook, and resulted in international press and popular support for women's rights activism in that context. As the images in Box 3.4 suggest, new media can thus be seen as a virtual space that supports and promotes organizing and activism in physical spaces.

Jennifer Fluri's (2006) analysis of the website of RAWA (the Revolutionary Association of the Women of Afghanistan) provides another example of how virtual spaces can be used to empower groups who are marginalized in physical space (in this case, through fundraising and creating a supportive international political network). However, Fluri's analysis also cautions readers that the same tactics used to meet these ends (documentation on the web of women's experiences of public violence) also reinforce problematic representations of Afghan women, as defined primarily by their status as victims of cultural and religious violence. This type of representation of women can thus serve to reinforce assumptions about the superiority of the West, while overlooking the agency and complex lives of women in post-colonial contexts. Similarly, while Katty Alhayek (2014) discusses the possibilities of online organizing for the benefit of Syrian refugee women during the conflict beginning in 2012, she also cautions that the same tools that are available to those resisting oppressive structures are available to governments, corporations, and other powerful

Spaces of culture and identity production

Box 3.4 Changing gender norms through new media activism

Figure 3.7 The logo of the Women2Drive movement. Reproduced courtesy of Carlos Latuff.

Figure 3.8 The logo of Feminist Frequency, a website created by Anita Sarkeesian, hosts a video web-series exploring issues related to gender and feminism. These videos are widely distributed on YouTube and used by media critics and gender educators. Reproduced courtesy of Feminist Frequency.

Activity

Take some time to visit the website of Feminist Frequency (www.feministfrequency.com) and explore the Women2Drive movement on social media websites. These two movements represent different ways in which feminists have used social media to create networks, form new alliances, and push for social change. Some critics argue, however, that as a neutral technology, social media is just as likely to perpetuate gendered norms as it is to challenge them. Think about your interactions with social media, including social networking sites such as Facebook, Twitter, Snapchat, YouTube, and Instagram, as well as other online blogs and forums such as Pinterest and Tinder. In what ways are norms regarding gender and sexuality reinforced? In what ways are they challenged? Consider issues such as avatar choice, who you interact with, what you read and write, and graphic representations. Given your experience, do you think that new media platforms constitute a space of liberation?

actors. As such, she urges us to recognize that the technology of the media itself is neutral, and that its effects are neither innately oppressive nor resistant in nature.

Given these insights, it is clear that new media and social media may also reproduce normative understandings of not only gender and sexuality, but also colonialist perspectives on nation, race, and cultural groups. Sophia Maalsen and Jessica McLean (2015), for instance, explore representations of gender and geographic space in an online music venue called triple j Unearthed (www.triplejunearthed.com), designed to promote and circulate new Australian music. They find that, in spite of the online venue and the explicit goal of promoting new music, the vast majority of the artists represented on the website's weekly popularity charts are men, while the women represented on the site tend to be 'providing support to, or supported by, masculine acts' (428). Thus, in these spaces, people and places are situated in relation to hierarchies of gender and geography, and as such, the democratizing possibilities of cyberspace are limited. Similarly, research on spatial media such as Open Street Map and the geosocial apps Girls Around Me and Wheretheladies.at, both of which are designed to use location data to identify social venues in physical space where women are present at a given moment, are strongly gendered. As researchers Agnieszka Leszczynski and Sarah Elwood (2015) have argued, these spatial media draw heavily on gendered and sexual norms to encode space in digital media in ways that, in turn, maintain and naturalize those norms.

In a similar way, Rashad Shabazz (2014) has explored the negotiation of gender roles and the media in spaces of hip-hop. Describing hip-hop as 'the location in which contemporary Black identity is played out,' Shabazz argues that:

> While women occupied the public spaces where hip-hop was cultivated, young men and boys used their access to the public space within Black geographies to not only counter their spatial disempowerment, but also to exclude women from these same spaces. Women's exclusion from hip-hop's productive spaces was the result of Black men's situating hip-hop's spatial resistance in their favor.
>
> (372)

Shabazz goes on in his analysis to contend that 'one of the most significant consequences is that it gives rise to distorted perceptions of women fostering narrow articulations of Black masculinity and establishing the ideological link between hegemonic masculinity and rap music' (377). In spite of this, Shabazz identified spaces such as 'safe free-play zones' and organizations such as Black Girls Rock! as sites that function to reshape the gendered geography of hip-hop by creating egalitarian spaces where women participate in ways that challenge norms of masculine dominance.

Overall, traditional and new media in many ways are like the other sites of cultural production discussed in this chapter, as they provide a vehicle for both the entrenchment of conventional understandings of gender and sexuality, as well as an opportunity to resist and redefine those discourses. Yet the possibilities of new media go beyond those conventional media, as it functions as a liminal space where the public and private may converge, where intimate details may be shared anonymously, and where new identities can be tried out and adopted or discarded. As such, it represents a unique opportunity for organizing, activism, and change relating not only to gender, but to other identity categories as well.

Conclusion

This chapter has examined three sites of the production of cultural meaning—the home, sites of consumption, and the media—using a feminist geographic perspective. In each site, gendered difference is produced through cultural processes in various ways. We have highlighted the ways in which other aspects of our identity (including race, class, and sexuality) are simultaneously created, challenged, and negotiated in these important spaces.

In the space of the home, for instance, we discussed how identities associated with alternative sexual identities can be both repressed (when the home is enforced as a heteronormative space) or expressed, as people create new meanings and models of the home in spaces over which they have control. In considering spaces of consumption, we highlighted how people create social meaning and reinforce their identities through shopping, leisure activities, and tourism. We provided the example of sex tourism as a complicated site in which gender differences are created and reproduced in the context of a racialized, post-colonial, geopolitical context. We also surveyed the ways in which gendered and spatial meanings are produced and circulated through the media. Increasingly, the media is an ever-present, powerful force in our society. With changing technologies and the advent of social media, our interaction with mediated messages about gender, difference, and power is constant and intense. Additionally, people of all types are increasingly becoming not only consumers, but also *producers* of media content, through venues such as social networking sites, blogs, and other social media platforms. This chapter has raised the question of what the effects of these changes mean for struggles over gender change and social justice.

While the three sites—the home, sites of consumption, and the media—have been considered separately in this chapter, they are all interconnected and function to support the production of identity and cultural meanings across not only these spaces, but other spaces as well. The home, for instance, functions as a critical site of consumption, including for activities such as shopping and leisure (Cox 2013). Similarly, as the examples of the geosocial apps described in the previous section indicate, the media has become critical to the way in which people understand and experience leisure activities. Increasingly, consumption and leisure activities that were once uncommon in the home are taking place in that space as we expand our interaction over and through social media. Each of these sites, both separately and together, thus function to produce gendered and sexualized meanings in our everyday lives.

Recommended reading

Blunt, Alison, and Robyn Dowling. 2006. *Home*. New York: Routledge.

Brickell, Katherine. 2012. '"Mapping and "Doing" Critical Geographies of Home.' *Progress in Human Geography* 36 (2): 225–44. doi:10.1177/0309132511418708.

Caudwell, Jayne, and Kath Browne. 2013. *Sexualities, Spaces and Leisure Studies*. New York: Routledge.

Cox, Rosie. 2013. 'House/Work: Home as a Space of Work and Consumption.' *Geography Compass* 7 (12): 821–31. doi:10.1111/gec3.12089.

Domosh, Mona, and Joni Seager. 2001. *Putting Women in Place: Feminist Geographers Make Sense of the World*. New York: Guilford Press.

Gregson, Nicky, Louise Crewe, and Kate Brooks. 2002. 'Shopping, Space, and Practice.' *Environment and Planning D: Society and Space* 20 (5): 597–617. doi:10.1068/d270t.

Johnston, Lynda. 2005. *Queering Tourism: Paradoxical Performances at Gay Pride Parades*. New York: Routledge.

Morrison, Carey-Ann. 2013. 'Homemaking in New Zealand: Thinking through the Mutually Constitutive Relationship between Domestic

Material Objects, Heterosexuality and Home.' *Gender, Place & Culture* 20 (4): 413–31. doi:10.1080/0966369X.2012.694358.

Rose, Gillian. 2016. 'Rethinking the Geographies of Cultural "Objects" through Digital Technologies Interface, Network and Friction.' *Progress in Human Geography* 40 (3): 334–51. doi:10.1177/0309132515580493.

Tan, Qian. 2013. 'Flirtatious Geographies: Clubs as Spaces for the Performance of Affective Heterosexualities.' *Gender, Place & Culture* 20 (6): 718–36. doi:10.1080/0966369X.2012.716403.

Waitt, Gordon, Kevin Markwell, and Andrew Gorman-Murray. 2008. 'Challenging Heteronormativity in Tourism Studies: Locating Progress.' *Progress in Human Geography* 32 (6): 781–800. doi:10.1177/0309132508089827.

Chapter 4

GENDERING THE RIGHT TO THE CITY

Risa Whitson

Introduction

Take Back the Night marches—held annually on hundreds of college campuses in North America and Europe, as well as in countries such as India, the Bahamas, and Japan—symbolize concerns about sexualized violence in, and women's lack of access to, public space. These marches, which began in the 1970s in response to violence against women in public spaces, also underscore the significance of a presence in public space for reclaiming rights and fostering a feeling of belonging. In recent years, other social movements have also highlighted the critical importance of access to, and control over, public spaces for functioning democratic societies, such as the Occupy movement and the Walk a Mile in Her Shoes rallies in North America. In addition to other locally-based movements present around the globe (for a list, see www.stopstreetharassment.org/resources/online/), transnational programs to address these issues have also been established (for instance, the UN Women's Safe Cities Global Initiative: www.unwomen.org/en/what-we-do/ending-violence-against-women/creating-safe-public-spaces).

The presence and proliferation of these movements, and others like them, highlight the ways in which the struggle over access to the city, representation in public space, and social belonging in the city continue to be gendered and sexualized. Feminist geographers have long focused on urban areas, and particularly urban public spaces, as sites of importance for examining social relationships and social justice issues. From the perspective of feminist geographers, the physical organization of the city, as well as access to and control over its spaces, are critical aspects of the construction of gendered and sexualized identities and inequitable power relations. In analyzing urban areas, feminist geographers have asked: How do these unique spaces play a role in constituting gendered meanings? Who has access to and control over these spaces, and to what end?

In this chapter, we address the gendered geographies of the city through the concept of the 'right to the city.' We discuss the ways in which feminist geographers have understood these rights (as well as cities themselves) to be gendered by exploring themes of public space and urban change. In particular, this chapter examines urban public space by considering the ways in which gendered public spaces affect people's behavior, identity, and mobility. Following that discussion, the chapter addresses urban change through an examination of the effects of neoliberal urban policies, including gentrification.

Gendering rights to the city

According to the United Nations Population Fund (UNFPA), we are now living in the 'Urban Millennium' (UNFPA 2007). Since 2008, over half of the world's population has lived in urban areas, and the number of urban dwellers is continuing to rise dramatically. While countries in the Global North have had a large percentage of residents living in urban areas since the middle of the 20th century, cities in the Global South continue to grow dramatically. Researchers estimate that over 90 percent of future urban growth will occur in the Global South, with the majority of that taking place in Africa and Asia. While cities have long held the promise of new economic, social, and political opportunities, for many people (including women and sexual minorities), this promise is not always the reality of today's urban centers. Although cities may offer greater freedoms from traditional norms, and increased possibilities for education and employment, high rates of urban growth are also associated with high levels of poverty, a lack of adequate infrastructure, and the increased marginalization and vulnerability of large portions of the population, particularly in the Global South. Around the globe, urban areas are at the center of debates over the provision and availability of services, as well as access to and participation in public space. Indeed, it is in these contexts that critical struggles are being waged over who belongs to 'the public,' whose voice should be heard, and who has the right to control the spaces around them.

During the 1990s, urban geographers began to discuss the importance of this **right to the city** in discussions surrounding access to resources, political rights, and citizenship. From this perspective, cities are not just backdrops against which social action and relationships occur. Rather, cities are places of encounter for everyday life, where rights can be realized and social belonging achieved. Drawing on the work of French philosopher Henri Lefebvre (1996), geographers thus argue that cities should constitute places of inclusion, participation, and access to resources. These ideas envision cities as central to the process of belonging to, and participating in, a political and social community.

According to Tovi Fenster (2005) and Mark Purcell (2002), the concept of the 'right to the city' has two primary components. First, it involves the **right to appropriate** urban space. This includes the right to 'live in, play in, work in, represent, characterize and occupy urban space in a particular city' (Fenster 2005, 219). For instance, the 'right to appropriate' the city means that people from all social groups can congregate in public spaces, such as parks, streets, and commercial areas, doing the activities that they enjoy without fear of police brutality, or economic or racial exclusion.

Second, the right to the city involves the **right to participation**. This means the ability to engage in meaningful participation in the decision making, political life, and management of the city. In other words, this is the right not to simply live in the city, but to participate in making decisions that shape the city according to one's own needs and desires. This requires that the people who live, work, and play in cities be involved in decision-making processes that affect and produce urban space.

The concept of the 'right to the city' has not been limited to academic conversations. As evidenced by the World Charter for the Right to the City, which emerged from the World Social Forum meetings in 2004 and 2005, the concept has become a tool for social organizing, more broadly. At a time when the majority of the world's population now lives in urban areas, and the rates of urban poverty continue to skyrocket, the idea of the right to the city provides a way of understanding human rights that is place-specific. This place-based conceptualization of human rights can serve as an anchor for struggles over social justice. Yet the gendered aspects of these rights to the city are often overlooked. While the United Nation's Sustainable Development Goal 11 focuses on the right to the city, it does not have an explicit gender component. Similarly, while Sustainable Development Goal 5 reinforces global interest in achieving gendered equality, it does not consider urban space as an explicit site of action to meet this goal (UN 2014).

It is not only in policy, but also in academic and theoretical discussion of the right to the city, that issues of difference (including, but not limited to, gender difference) have been insufficiently explored. According to Fenster (2005), Lefebvre's understanding of the right to the city, which lays the foundation for much of the thinking on this topic in urban geography, does not 'challenge any type of power relations (ethnic, national, cultural) let alone gendered power relations as dictating and affecting the possibilities to realize the right to use and the right to participate in urban life' (219). This is to say that a generic understanding of the right to the city does not take into consideration how some people (such as men, the wealthy, the highly educated, ethnic and racial majorities) are more free both to use the city as they would like, and to participate in decision making regarding these areas (Box 4.1). Moreover, as Linda Peake and Martina Rieker (2013) argue, 'achingly little attention has been given specifically to the roles women are playing in these [urban] processes and the social, economic, and political implications of these new urbanisms and spatialities

Gendering the right to the city

Chapter 4

Box 4.1 Who has the right to your city?

Figure 4.1 A march to protest police brutality during the second week of Occupy Wall Street in New York City. Photo taken on September 30, 2011. Reproduced courtesy of David Shankbone.

Figure 4.2 Black Lives Matter activists protesting in Minneapolis, Minnesota. Photo taken on November 15, 2015. Reproduced courtesy of Tony Webster.

The two images above illustrate moments in which two groups—the Occupy movement and the Black Lives Matter movement—have appropriated public space to confront social, economic, and political injustices. In moving their protests into public space, both groups also raised questions about who belongs in, and controls the use of, urban public spaces.

Whitson

Activity

Consider the public and semi-public places in which you live, work, and play. Who has the right to appropriate these spaces? That is, who has the right to use them, and for what purposes? As you answer this question, consider a variety of social groups: women and men, sexual minorities and straight people, affluent and low-income people, people of color and white people. Who would be welcomed and who would be viewed with suspicion? Would some activities be considered allowed for certain groups, and viewed with suspicion for other groups? How is people's presence and behavior constrained through law or custom? The 2014 shooting of Tamir Rice and the 2012 shooting of Trayvon Martin, for instance, highlight the way in which race and class work together to restrict the access of lower-income African Americans to public space. Consider also who has the right to participate in the production of public spaces. In other words, who makes the decisions about these spaces? Who creates the rules about what public spaces should be used for, and by whom?

for them' (12). As a result, feminist geographers have worked to explore the rights of women and other groups to appropriate urban space and to participate directly and meaningfully in its production (Fincher and Jacobs 1998).

Gendered public space in an urban context

For those interested in the right to the city, consideration of access to, control over, and a feeling of belonging in public spaces is one of the areas of greatest concern. In the context of democratic societies, **public spaces** (such as parks, streets, plazas, and other spaces of supposedly open access) are especially important, as they provide an opportunity for interaction with those different from us outside of the direct control of the state or economic actors (such as businesses) (Bondi and Rose 2003). Equally importantly, they provide a space for people 'to go and just be' (Mitchell 1995, 110). Not only is this important for the organization of political movements, but a presence in public space also allows for different groups, including marginalized groups, to become visible and legitimate members of the broader public. As such, public spaces are a key site in claiming the right to appropriate and participate in urban space (Secor 2004).

While the ideal of public space is an inclusive concept, in practice it is, and has always been, exclusionary, limiting access to women, people of color, linguistic minorities, people who identify as LGBTQ, working-class and poor people, the elderly and children, and the disabled (Young 1989). According to Liz

Gendering the right to the city

Chapter 4

Figure 4.3 Central Park in New York City. Parks are an iconic example of public spaces, where people may have the possibility of encountering those different from them. This image suggests, however, that while no formal restrictions may exist prohibiting the use of public space by certain groups, these spaces may nonetheless be used by people who are fairly homogenous in terms of class and race. Reproduced courtesy of Woody Wood.

Bondi and Damaris Rose (2003), 'public space is understood to be constituted by impositions, negotiations and contestations over which groups comprise the public that has access to these spaces, for what purposes these spaces are used, and what visions of society urban public space embraces, enforces, produces and promotes' (235). Feminist geographers in particular have argued that public spaces are constructed as masculine and heteronormative, effectively preventing the access, participation, and belonging of women and sexual minorities.

A number of feminist geographers have chronicled the historical development of gendered public space in the context of Western Europe and North America, beginning with the importance of the separation of spheres to the masculinization of public spaces (discussed in Chapter 3). By the early 20th century, as Daphne Spain (2014) recounts, separate spaces were strongly entrenched in the city:

> Department stores, settlement houses, and women's clubs were gendered spaces in the industrializing cities, as were saloons, restaurants, and business clubs. They gave men and women different destinations in the city. No signs were needed to indicate which sex belonged in which space; it was simply understood who belonged where. Thus, the idea of separate spheres became encoded within the public spaces of the city.
>
> (584)

Gender was not the only category that was used to organize access to public space. As discussed in Chapter 6, race has also been a way of explicitly organizing urban space, in both the US and other contexts. While spatial segregation based on race was encoded in law, and often included visual markers such as signs, gendered spatial segregation was often less codified, but nonetheless had strong effects on people's access to, and use of, urban spaces.

In many parts of the Global South, patterns of public and private space have also historically mapped onto understandings of the masculine and feminine. This conceptualization can also extend to understandings of space as forbidden and permitted, as is evident in Bedouin communities in Southern Israel. According to Tovi Fenster (1999), in this context, rather than spaces being categorized absolutely as 'public' or 'private,' spaces take on different meanings for different individuals at different times. Thus, the private space of the home may typically be 'permitted' for women's space, but once a male stranger enters this space, it becomes 'forbidden' to the women who live in the house, who must then seek out more modest permitted space or use the 'portable seclusion' of the veil (Papanek and Minault 1982). Yet while Muslim women are often perceived as uniquely restricted in their use of public space, this is not necessarily the case. Rather, women in all contexts must negotiate issues of risk, cultural norms, and gendered household responsibilities in public spaces. These spatial mappings of public/private and forbidden/permitted are critical in understanding people's experience of the city, as they establish a spatial system of organization, within which people of all genders negotiate and inhabit the city. This in turn provides a context for the possibilities and constraints available to people of different genders with regards to urban participation and appropriation.

Gendering the right to the city

4

For those of us living in contemporary urban areas, the world that Spain (2014) describes above may seem to have little in common with our urban experiences today. Yet in contemporary cities, there continues to be gendered differences in the use, experience, and control of public space. That is, for people of different gender and sexual identities, there are distinct patterns of presence and mobility, behavioral norms, feelings of belonging, and ability to control and regulate the activities and behaviors of others in public spaces.

These differences may be manifested in various ways. In some places, customs and religious beliefs continue to exclude women from public spaces. For instance, in the context of Saudi Arabia, while women-only public spaces exist, interaction in public space between the sexes continues to be a subject of intense debate. In this context, Saudi women may also find it necessary to be accompanied by a male relative to enter public space, or may restrict their movement to women-only public spaces (van Geel 2016).

In other places, while no laws may exist to prohibit women's entrance into public space, gendered behavioral norms nonetheless continue to persist. Chapters 1 and 2 examine how these norms can affect one's experience of, and interaction with, public spaces, including by affecting one's dress, mobility, and leisure activities. For example, in some contexts it is less common or unacceptable for women to walk alone at night, be present in public space without a defined reason (that is, just 'hanging out'), or engage in public leisure behaviors considered to be 'masculine,' such as drinking and smoking. While these norms may be enforced through social expectations or the threat of violence (implicit or explicit), they are also frequently incorporated into gendered performances of femininity. As a result, women often internalize these gendered norms of behavior in public space in such a way that the norms do not need to be externally imposed. This means that women will police themselves to conform to these norms, thus removing any need for formal rules regarding what women can or cannot do in these spaces. Janaki Abraham (2010) argues that 'girls and women learn how to produce "respectability" through their use of spaces and clothing, while simultaneously testing the boundaries of this control' (201). These gendered performances are learned at an early age, as girls continue to have different experiences of mobility and activity in public spaces than boys (Figure 4.4).

Public spaces are also gendered through images, such as outdoor advertising. These images often work to sexualize women and present them as adornment in public space, rather than as the creators or authors of such spaces. Therefore, their inclusion in public space through visual imagery supports

Chapter 4 Whitson

Figure 4.4 Hashemite Plaza in Amman, Jordan. While public spaces may be open and available to women, internalized gender norms continue to structure behavior in public spaces in all geographic contexts. Reproduced courtesy of Woody Wood.

women's subordinate status (as visual and sexual objects) and reinforces the masculine nature of these spaces (Rosewarne 2005).

In exploring the experiences of gendered individuals in public space, however, it is important to recognize the effect of not only gender on experiences of public space, but also other aspects of identity, including class, race, sexual identity, age, ability, and nationality. These factors also strongly impact individuals' ability to claim rights to participate in, and appropriate, public space. Further, because gender roles and expectations vary from one social group to another, not only do women in different places have different experiences of the city, but women of different races, ethnicities, classes, ages, and nationalities will also experience and subsequently produce public space in different ways. In spite of this complexity, gendered experiences of public space lead to, and are subsequently reinforced by, gender differences in feelings and symbolic meaning associated with public spaces.

Gendering the right to the city

Public space and the geography of fear

One of the most prominent themes that feminist geographers have explored in relation to public space has been women's experience of fear. In a groundbreaking article on this theme entitled 'The Geography of Women's Fear,' feminist geographer Gill Valentine (1989) argues that women's experience of fear in public spaces affects their use of these spaces in terms of engagement and mobility, and often results in them seeking 'protection' from men as they enter into public space (for example, by asking a man to walk them home). Valentine argues that not only are these behaviors an expression of women's disadvantaged position within a patriarchal social system, but that they also in turn maintain and perpetuate male dominance through reinforcing men's control over public space. Thus, Valentine concludes that women's 'inhibited use and occupation of public space is therefore a spatial expression of patriarchy' (389).

A number of other geographers have subsequently explored gendered experiences of fear in public space, examining the reasons for, and effects of, fear on individuals and social groups (England and Simon 2010). This body of work has concentrated on the various factors associated with fear in public spaces, how the experience of fear varies between different groups, and how men and women react to these fears. Important to understanding this subject is realizing that geographies of risk do not correlate with geographies of fear. That is, men are more likely than women to be victims of violence in public space, yet report lower levels of fear. At the same time, women's fearfulness in public spaces is ubiquitous and not closely related to risk of crime, as women are actually much more likely to be victimized in private spaces. While certain types of places are associated with women's fear (such as dark, isolated spaces, and local public spaces such as parks, paths, transit stations, and shopping centers), it is not these built environments that create the fear, but rather gendered social power relations—resulting in women's subordinate social, economic, and political status—that produce social vulnerability. This is consistent with research that suggests that fear is connected to social well-being, such that those who feel most vulnerable in society overall will tend to be the most fearful (Koskela 1997).

Women's experience of fear in public spaces has important consequences that affect social interactions and women's use of space and quality of life. In particular, women's strategies in the face of fear of violence cause them to limit their movements, avoid certain areas of the city at certain times of day, restrict unaccompanied traveling, and keep to their homes (Box 4.2). These strategies reinforce social norms that associate the masculine with public space, normalize women's fear of such spaces, and perpetuate the discourses that function to produce this fear. In her research on women's experiences in public spaces in Mumbai, India, Shilpa Phadke (2005) explores how safety comes to take the form of surveillance for women. She states:

> To be safe is not only to be policed externally, but also to internalize the panoptic gaze that demands normative behavior denoting a sexuality under control. Safety in the form of surveillance might in fact create a context within which women, too, could be more carefully monitored by families, communities and, in a more general sense, by the conservative sections of society.
>
> (55)

Nonetheless, women also respond to these geographies of fear not only through avoidance, but also through boldness and the production of spaces where they feel safe. Just as fear is an expression of social vulnerability, women's boldness in public space works to open up these spaces for increased appropriation by women. As Hille Koskela (1997) argues:

> The fact that some women are bold and confident shows that women are not only passively experiencing space but actively take part in producing it. They reclaim space for themselves, not only through single occasions such as 'Take Back the Night' marches, but through everyday practices and routinised uses of space. Their everyday spatial practices can be seen as practices of resistance. By daring to go out – by their very presence in [the] urban sphere – women produce space that is more available for other women. Spatial confidence is a manifestation of power. Walking in the street can be seen as a political act: women 'write themselves onto the street.'
>
> (316)

Women's feelings of fear and safety are thus intertwined in public spaces, making women's everyday choices to engage in public space an expression of their ability to live with and move beyond feelings of fear (Panelli et al. 2005; Phadke 2005).

These experiences of resistance and agency by women in urban public space are also reflected in the contemporary activism around this issue, including by the organizations discussed in the introduction to this chapter and Chapter 1,

Gendering the right to the city

Box 4.2 Fear in public space

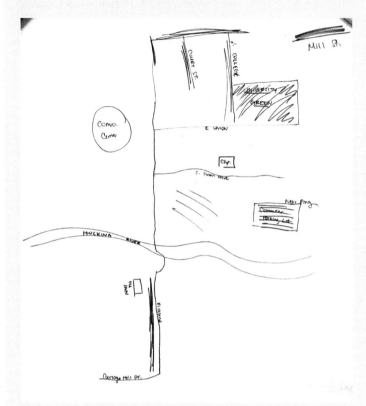

Figure 4.5 A mental map of spaces of fear produced by female college students. In this map, hashed areas indicate spaces of fear. Reproduced courtesy of Maria Panaccione.

Figure 4.6 A mental map of spaces of fear produced by female college students. In this map, students illustrated spaces where they felt fearful, including images of people and activities within the spaces. Reproduced courtesy of Maria Panaccione.

Figures 4.5 and 4.6 were produced as part of a mental mapping exercise with female college students that was designed to illustrate what they perceive as spaces of fear on campus. In Figure 4.5, students clearly identify areas that are unsafe with hashed marking. In Figure 4.6, particular bodies, social relations, and activities are identified that make the women feel less safe.

Activity

In groups or as individuals, consider the ways in which you move through public spaces in your everyday life. What would your mental map of spaces of fear look like? Are there areas that you avoid? Do you take different routes at different times of the day? Are you careful to walk in groups, or do you feel comfortable walking alone? Do you have strategies planned for circumstances in which you might feel threatened? Have you ever decided not to do something that you wanted to do, as a result of these fears (for example, not taking a job with late hours, not going to the library at night, or not walking home alone from a party when you were ready to leave)? Consider how your experience with fear in public space is a result of your sex, gender identity, sexual orientation, class, race, or other aspect of your identity. Think about how your experience is similar to, or different from, people in other social groups. What conclusions can you draw from your personal experiences and those of your classmates about patterns of fear in public space and their effects on different groups?

such as SlutWalk, Take Back the Night, Walk a Mile in Her Shoes, and the UN Women's Safe Cities Global Initiative. In some contexts, such as Mexico City, Jakarta, Rio de Janeiro, and Tokyo, governments are also working with local activists to organize women-only taxis, buses, and train cars in order to create safe ways of moving throughout cities. In other contexts, women are organizing through grassroots initiatives that blend online and physical activism, such as Hollaback! (www.ihollaback.org), an organization with a website and mobile app which documents harassment against women in public space, and encourages women to use mobile and social media to reclaim these spaces, and for mutual support. Similarly, the victimization of women in the Tahrir Square protests in Cairo, Egypt during and following the Arab Spring uprisings in 2012 and 2013 has also given rise to organized resistance movements focusing on the right to public space. One such organization is HarassMap (www.harassmap.org/en), which focuses on supporting women's right to the city, and their presence in public space, by in-person individual and group interventions designed to address and end gendered violence in public areas.

A third innovative initiative, Why Loiter (found at #whyloiter on Twitter and Facebook and whyloiter.blogspot.com), has begun in Mumbai, India. This initiative draws on the work of authors Shilpa Phadke, Sameera Khan, and Shilpa Ranade (2011) and encourages women to intentionally loiter in public spaces in order to challenge the social restrictions on women's movement in public space.

Increasingly, researchers have also begun to explore the connections between masculinity, fear, and public space. This research suggests that while women's fear in public space is problematized, men's fear is normalized and, as such, often remains overlooked and unexamined (Day et al. 2003). Yet as a result of the way in which normative masculinities are constructed, men both experience and react to fear differently than women in public spaces. In particular, men's experience of fear in public space is often centered on concerns around physical confrontations that require them to 'prove themselves' through fighting (Day et al. 2003). Yet rather than engage in spatial strategies of avoidance that are common among women, researchers argue that men 'perform fearlessness' through strategies designed to maintain control over their environment (Brownlow 2005). These strategies 'assert authority, achieve power, and sustain patterns and modes of access to and control over the public sphere that are generally unavailable to women' (Brownlow 2005, 584). As such, men's experiences of fear, as well as their response to women's fearfulness, work to reinforce their presence in, and control over, public spaces rather than subvert them.

As a result of the connection between social vulnerability and affect, experiences of fear in public space often reflect intersectional identities such as race, class, sexual identity, and age. For women, while fear in public space is something that cuts across identity categories, experiences of fear are very classed and racialized. This is true both in the sense that the places about which, and people of whom, they feel fearful are affected by class and race (for example, white women are more fearful of non-white, poor 'others'), and in the sense that these factors affect the form the fear takes (for example, fear of sexual assault, fear of racial hostility, a feeling of being 'out of place,' etc.) Thus, for women of color and poor women, fear in public space also may include a fear of racism and class prejudice, and may be heightened in majority white or affluent areas. Accounts of experiences of African American men in public space also vividly illustrate the way in which race informs men's feelings of fearfulness in public space. For instance, Ta-Nehisi Coates' (2015) account of his experience as a Black man in the contemporary US describes the ways in which his actions and interactions in public space have been informed from the time of his youth by a fear of police violence, surveillance, and institutionalized racism. This experience of fear goes beyond that of a generalized concern with 'confrontation' as described above. Moreover, the research of Kristen Day (2006) highlights the way in which young men of color, in particular Hispanic men, experience and understand being feared by others as a result of their race. Additionally, for transmen and transwomen, concerns with anti-trans violence and being 'outed' also create distinct experiences of fear in public spaces (Doan 2009). As such, age, race, class, sexual identity, and gender presentation should be viewed not simply as characteristics of individuals, but as social processes that help to explain patterns of fear as a result of social power (Pain 2001).

Gendered experiences of neoliberal cities

In addition to considering gendered experiences of urban public space and their implications, feminist geographers have also been concerned with how changes in urban governance and economic policy have gendered implications. In particular, since the 1980s, feminist scholarship has focused on the role of neoliberal urban policies in the lives of women and LGBTQ communities. **Neoliberalism** refers to a philosophy of governance that prioritizes economic liberalization, thus favoring competitive, free-market-based systems over direct state intervention. As such, it is characterized both by privatization and deregulation, as well as the 'proliferation of regulatory systems through which quasi-market mechanisms have been extended into an ever-widening range of activities' resulting in 'market-style competition for public funds' (Bondi and Laurie 2005, 3). In the context of urban policy, this has resulted in not only the privatization of local public services, but also the restructuring of urban housing markets through the elimination of public and low-rent housing, the destruction of low-income neighborhoods, and the introduction of gated communities. It has also led to entrepreneurial urban revitalization programs, often resulting in gentrification and discriminatory surveillance and social control measures (Peck et al. 2009). This is because neoliberal policies that seek to reinvest in the city through the development of high-end consumer and business services catering to the affluent are accompanied by 'zero-tolerance' policies directed toward the poor, the homeless, prostitutes, and other urban residents who are viewed as unsightly in this new neoliberal landscape. As such, marginal groups are systematically pushed out of these spaces as 'they simply cannot afford to participate in leisure rituals that revolve around designer shopping and the consumption of *caffè latte*' (Hubbard 2004, 669). Increasingly, this is occurring both in the Global North and the Global South.

Social reproduction and the city

While commonly perceived as gender-neutral, neoliberal urban policy in fact has gendered dimensions and consequences. Because the city provides access to the basic needs critical for daily life (such as housing, food, transportation, and work), it serves as a key site for **social reproduction**. Cindi Katz (2001) defines social reproduction as 'the biological reproduction of the labor force, both generationally and on a daily basis' (711). As a result of the continued central role that women play in social reproductive processes, they are disproportionately impacted by neoliberal urban policies (Box 4.3). Some of the policies that particularly impact women include the privatization of social services, the destruction of low-income housing, and urban revitalization that marginalizes those who cannot conform to the norms of elite, conspicuous consumption. This is especially the case for low-income women and women of color. As Matthew Desmond (2012) documents in the context of Milwaukee, for instance, African American women are the group most likely to be evicted and most disadvantaged by eviction. Because women, on average, have lower incomes, a greater number of dependents, and more resource-poor networks than their male counterparts, processes resulting in the removal of low-income and subsidized housing affect them more frequently and more intensely. Lack of adequate public transport can also adversely affect women, people of color, and low-income groups, as they have, on average, less access to privately owned vehicles. In contexts in which women are prohibited or discouraged from driving by law or custom, access to adequate public transport can become an even more important issue.

Similarly, while women are most likely to be responsible for social reproduction, they have the least access to urban housing and land, formal work that is well paid, and other critical public services such as sanitation and water (Chant 2013). This is especially true in the Global South, where these challenges pose urgent problems for low-income women (Chant and McIlwaine 2015). As neoliberal urban development projects increasingly seek to re-purpose urban spaces for high-rent activities such as upscale malls or condominiums, low-income housing or squatter settlements may be demolished and informal street vendors relocated to the margins of town, where they have less access to

Box 4.3 Access to infrastructure for social reproduction in Madrid

One way in which feminist geographers have explored the effects of neoliberal urban policy on women's lives is through the use of GIS, or Geographic Information Systems, to visualize and analyze women's access to resources needed for social reproduction in urban areas (Pavlovskaya and St. Martin 2007). In this example, feminist geographers Jose Carpio-Pinedo, Sonia De Gregorio Hurtado, and Ines Sanchez de Madariaga engage in counter-mapping to visualize silenced narratives and reveal spatial inequities in access to critical infrastructure for social reproduction and everyday life. Figure 4.7 illustrates access to pharmacies, sports centers, and health centers in Madrid by highlighting walking distances to these facilities. Figure 4.8 maps a global score for all infrastructures for everyday life to help visualize general levels of spatial inequality in access to resources needed for social reproduction. While these maps are not explicitly gendered, because women are more likely to be responsible for social reproduction they will be differentially affected by this spatial inequality. Additionally, to the extent that women, immigrants, people of color, and low-income groups rely disproportionately on public transportation, access to walking-distance infrastructure for social reproduction will clearly reflect intersectional identities. As noted above, neoliberal urban policies can affect this access, as public transportation systems and the resources themselves may be run, funded, or subsidized by the state.

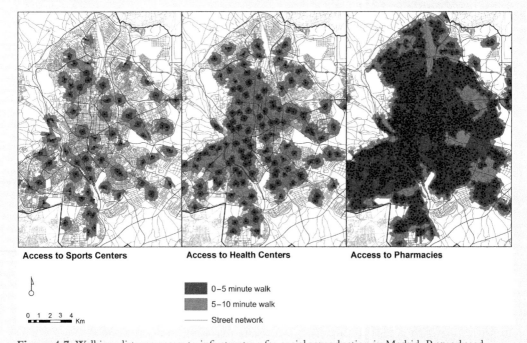

Figure 4.7 Walking-distance access to infrastructure for social reproduction in Madrid. Reproduced courtesy of Jose Carpio-Pinedo, Sonia De Gregorio Hurtado, and Ines Sanchez de Madariaga.

Gendering the right to the city

4 Chapter

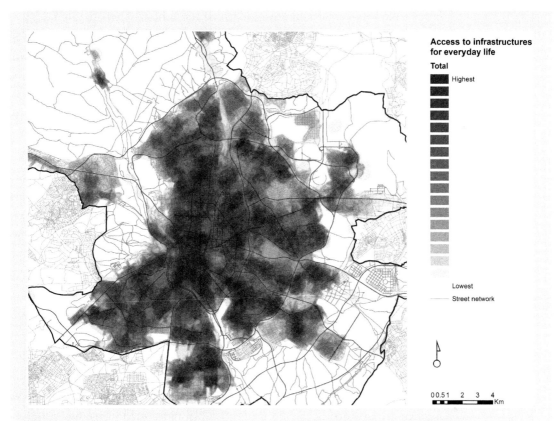

Figure 4.8 Global measure of access to infrastructure for social reproduction in Madrid. Reproduced courtesy of Jose Carpio-Pinedo, Sonia De Gregorio Hurtado, and Ines Sanchez de Madariaga.

Activity

Consider the infrastructure for social reproduction in your town (such as health centers, doctor's offices, grocery stores, day care centers, and parks). What groups of people have greater access to these infrastructures? Has access changed over time?

customers and suppliers. The research of Sapana Doshi (2013), for example, examines the experience of slum resettlement processes that are driven by, and beneficial for, neoliberal urban redevelopment processes. In analyzing women's activism surrounding low-income housing in Mumbai, India, Doshi details the gendered experience of the loss of housing and forced resettlement due to slum clearance. Kate Swanson (2007) also highlights the ways in which such urban regeneration policies have not only gendered but also racialized consequences. She argues that the removal of informal workers, beggars, and street children in an attempt to modernize and renovate Ecuadoran cities is in part to make them tourist-friendly but also fundamentally about 'whitening' the landscape of the city. Cindi Katz (1998) argues that, in this way, 'it becomes harder to find the everyday traces of lives spent among the monuments, as many are simply erased or removed from view' (42).

In addition to access to housing, transport, and income-generating opportunities, access to sanitation is another important element of social reproduction that is challenged by the increasing neoliberalization of global cities. As is also discussed in Chapters 5 and 7, in contexts in which neoliberal redevelopment is prioritized above the provision of basic needs for residents, a lack of access to sanitation can crucially shape women's experiences. While both men and women are affected by inadequate sanitation infrastructure, because of women's biological needs throughout their life course (including menstruation, pregnancy, and incontinence in old age) and because of their social responsibilities, a lack of access to sanitation infrastructure—especially working and usable toilets—can impact every aspect of low-income women's lives (Greed 2015; Penner 2013). This includes their access to work and education, as well as their ability to engage in basic activities necessary to social reproduction, such as cooking, cleaning clothes and dishes, and washing (McFarlane et al. 2014). Yet, as in the case of violence against women in public spaces, women are organizing around issues of social reproduction around the globe to address these issues (Box 4.4).

Box 4.4 The Right to Pee campaign

As in other growing urban areas in the Global South, over 50 percent of homes in urban India do not have toilets. The research of Colin McFarlane et al. (2014) describes the results of the lack of access to basic sanitation in the context of Mumbai. According to this research, this lack of access can result in complex gendered negotiations surrounding public toilets, leading to urination and defecation in open space. They argue:

> The geography of open defecation is at once spatial and temporal and generally structured in this way: Children will use the lane outside houses or the roadside, an often dangerous space that sometimes leads to injury and even death due to reversing municipal garbage trucks on the main road; men will use the *kabrastan* (graveyard) or *maidan* (open

ground) nearby the neighborhood; and women—sometimes with children and often in groups—will go further afield and use the garbage ground. Men, too, sometimes use the garbage ground, although they tend to use the lower edges that run along the large open stormwater drain, whereas women climb the garbage heap that stretches a few stories high.

(1006)

While the lack of access to sanitation thus endangers all of those who experience it, age and gender structure the experience as well. In particular, McFarlane et al. argue that, for women, needing to defecate in the more remote areas of garbage dumps may result in harassment and assault. For many other people who have access to public, shared, or private toilets within their home, the lack of public toilets within the city as a whole presents a different problem, particularly for low-income residents who cannot pay to use restrooms in privatized public spaces. These problems take a gendered form for women, whose embodied experiences as reproducers and caretakers result in greater health risks to themselves and their children. These challenges are compounded by the limited number of public toilets for women: New Delhi, for example, has 1,534 public toilets for men and only 132 for women. Moreover, higher fees for the use of paid public toilets for women, as well as stigmas against menstruation that result in women being turned away from public restrooms controlled by male attendants, also limit women's access to safe conditions of sanitation in Indian cities (Yardley 2012).

In response to this situation, a coalition of NGOs and grassroots activists, initiated by Mumbai-based activist Mumtaz Shaikh, began the Right to Pee campaign in 2011, with the aim of pressuring local governments to provide adequate, safe, free public toilets for women. 'Mumbai has 3,000 free public urinals for men whereas women do not have a single one . . . Do planners of towns think we are less human?' asks Shaikh (2015). Since its inception, the Right to Pee campaign has resulted in the creation of hundreds of public toilets for women, and the coalition continues fighting for more.

Gentrification in urban spaces

The relationship between gender and **gentrification** has been one area of particular interest to feminist geographers since the early 1990s. Gentrification is a critical feature of neoliberal urbanization and involves the revitalization of older, deteriorated urban areas. This process is often accompanied by the displacement of low-income residents by new middle- and upper-class residents and businesses. Women's increased participation in the workforce and changing family structures are two of the factors that have enabled gentrification to occur. Indeed, middle- and upper-income women are an important gentrifying constituent and have benefitted from increased access to what are perceived as 'safe' urban areas. Similarly, as will be discussed below, the development of gay neighborhoods has also played a role in the process of gentrification. However, not all women or members of the LGBTQ community benefit from this process of urban and neighborhood change. Rather, low-income individuals, including women and sexual minorities, are forced to bear the brunt of higher property values, increased code enforcement, and shortages of low-income housing that accompany the process of gentrification.

Feminist geographers have explored not only the differential gendered effects of gentrification, but also how gender itself is an organizing logic behind the gentrification process. Marguerite van den Berg (2012, 2013) uses the term 'genderfication' to describe 'the production of space for different gender relations,' which is a critical part of the gentrification process. She identifies the way in which women and families function as 'gentrification pioneers' in order to produce a space that is seen as family- and child-friendly and, as such, promotes urban regeneration. She also argues that in the context of Rotterdam (the Netherlands), state-led gentrification is being implemented through re-gendering the city itself, from working-class and masculine to an inviting feminine city associated with leisure, creative industry, and middle-class family dwellings: 'Rotterdam may have been a "muscleman" in the past, but . . . is now a creative temptress. The "city as woman" is a tempting femme fatale, luring new people with its virtues: "stylish, powerful, creative, inspiring, sparkling and full of surprises"' (van den Berg 2012, 161). Feminist geographer Leslie Kern (2010) similarly argues that gender narratives shape and help to justify new-build gentrification, in her study of condominium development in Toronto. In her analysis of the ways in which developers target women for urban condominiums, Kern argues:

Gendering the right to the city

Home ownership, condominium security, and the interconnectivity and freedom of choice found in city life are held up as ways for women to achieve independence and autonomy. Women's freedom in the central city is thus reified as a symbol of successful revitalization.

(375)

Overall, the presence of women may both promote gentrification through the suggestion that a neighborhood is safe and family-friendly, and it may also serve as a signal that revitalization has been successful (Kern 2013).

The process of gentrification is not only gendered, but is also sexualized, as it connects to sociospatial processes linked closely with sexual identification. In particular, early gentrification in many large urban areas in North America often occurred in areas that had large residential concentrations of gay men, as well as businesses catering to this community. During the 1970s and 1980s, primarily white, professional, gay men were, in many cases, the earliest gentrifiers as, constrained by social prejudice to live in deteriorating areas of the city, they used personal finances to support neighborhood revitalization. Yet, as with the presence of professional women discussed above, concentrations of gay men brought more than financial reinvestment in these older urban areas, as these neighborhoods became associated with alternative lifestyles, cultural opportunities, consumer amenities, and safe living spaces. Interestingly, however, while these **queer spaces** were among the earliest areas to be gentrified in the 1970s and 1980s, more recent neoliberal urban policy has resulted in continued redevelopment of these areas, such that gay men increasingly find themselves priced out of these communities, which have become expensive, trendy neighborhoods, appealing to wealthy investors, as well as local development, tourism, and urban planning offices. As a result, in some cities capital-intensive gentrification has resulted in the dispersal of LGBTQ residents, the closure of LGBTQ businesses, a loss of tolerance for the LGBTQ community, and a lack of participation by LGBTQ residents in the urban planning process. In addition to this, even in neighborhoods that remain enclaves for the gay community, members of other LGBTQ groups do not necessarily find spaces for themselves. While there has been some lesbian presence in many of these communities, the development of lesbian enclaves has been limited by women's lower incomes, their lack of access to capital, and concerns surrounding sexualized violence (Doan and Higgins 2009). Moreover, as in other parts of the city, these spaces remain strongly normative in other ways, such that queer women and gender-queer individuals may perceive them 'as supporting a certain form of mainstream gay culture that privileges and promotes certain sexualized, gendered, classed and racialized identities' (Nash 2013, 204).

While some women and members of the LGBTQ community have thus benefitted from gentrification, as it has opened up access to participate in, and appropriate, the city in new ways, this is not the case for all individuals. As the research of Caitlin Cahill (2006) indicates, experiences of gentrification are predicated upon class and race privilege. Cahill's research documents the process of gentrification from the perspective of young, working-class women of color, whose neighborhoods are experiencing the disinvestment and change that gentrification signifies. She draws on the results of her participatory research with young women in New York's Lower East Side to argue that it is precisely the stereotypical representations of low-income women of color as promiscuous, lazy, uneducated, and a burden to society that 'play an important role in facilitating processes of gentrification' (346).

Similarly, Melissa Wright's (2014) research underscores the impacts of neoliberal policies resulting in gentrification on low-income, working-class women in Ciudad Juarez, a city that was built by and for the working-class migrant women who worked in the city's factories. She describes how, during the 1990s, as part of an urban revitalization plan that sought to gentrify the city center, city officials attempted the removal of the sex workers and other low-income women who frequented urban public spaces. In Wright's words:

> They presented their plan as a way to 'clean up' the *centro* and 'reestablish' it as a place for upstanding families. That the downtown had never, in its entire history, been this kind of respectable place did not matter in their presentation of this formerly innocent past. Instead, the elites peddled their gentrification plan around a politics of forgetting about the working poor women and men who had established the *centro* as a place of commerce and tourism, where people of modest means worked, lived and relaxed. The development plan, in other words, presented the sex workers not as the city's original labor force, as is well documented in the city's historical record, but instead as invaders who had destroyed a formerly pristine social fabric.
> (7)

Together, the work of van den Berg, Kern, Cahill, and Wright highlights the importance of intersectionality to women's experience of neoliberal urban change. While high-income women may have increased access to 'safe' spaces (including work spaces, high-end residences, and sanitized public spaces of consumption) within urban areas, these spaces are increasingly created through the removal of those women who do not fit the neoliberal norm, such as prostitutes, the homeless, and the poor.

Despite these obstacles, marginalized groups who are affected by neoliberal urban policy are actively working to re-create urban spaces according to a different vision of urban space. Organizations such as Red Habitat (www.red-habitat.org), which functions in Bolivia, engage women in not only envisioning and lobbying around the theme of sustainable low-income housing, but also involve women in planning and building social housing that meets their needs. Similarly, in the context of the US, Roberta Feldman and Susan Stall (2004) document the decades-long activism of African American women to counter disinvestment in public housing in Chicago. Such activism is the on-the-ground expression of the struggle for the right to the city, as women appropriate, create, and participate in the cities they envision.

Liberatory potential of urban public space

In spite of women's experiences of fear and subsequent limitations on their behavior in public spaces, and despite the negative repercussions of increased neoliberalization, feminist geographers have also understood the city as an important venue for promoting the rights of marginalized groups. This is because the city is a unique place of possibility for the expression of identities that are outside of the mainstream. An early expression of this argument is present in Elizabeth Wilson's book *The Sphinx in the City* (1991), which explores the potential that cities hold for liberation, choice, possibility, and freedom. The city's liberatory potential is based on the possibility of increased encounters with others that cities, and particularly public spaces, offer. Cities may also be understood as 'modern' places in comparison to rural or provincial areas, allowing people to challenge norms and adopt identities that would otherwise not be open to them.

One example of this is provided in the research of Carla Giddings and Alice Hovorka (2010), exploring the gender performances of young women in Gaborone, the capital city of Botswana. This research shows that young men and women associate the city with modern gender roles, and as such, when moving to the city, may perform gender in ways that challenge traditional roles. The possibility of challenging gendered and sexualized norms in cities is often connected to the increased possibilities that cities provide for interaction and exchange with unfamiliar ideas, groups, and individuals. Using the example of Pakistani Muslim women's experiences of living in urban areas in the UK, Robina Mohammad (2013) argues that multiethnic, multicultural cities represent new 'shared spaces of interaction and exchange to open up possibilities for experimentation with cultural styles promoting new cultural practice, patterns of consumption, and alternative repertoires for a cosmopolitan Muslim identity informing feminine Muslim spatialities' (1819). As a result of this, she argues that the city provides young women with new ways of 'remaking the self' that may challenge not only traditional cultures in their homes and communities, but also work to normalize visible Muslim identities in public spaces (Figure 4.9).

Cities may be liberating not only for women, but also for sexual minorities (Huang and Yeo 2008). Places that disrupt the dominance of heteronormative space are increasingly found in urban areas. The best example of this is the establishment of **gay enclaves** (discussed above in association with gentrification) in cities in the Global North from the 1960s to the 1980s, through

Figure 4.9 Jardin du Luxembourg in Paris, France. Feminist geographers argue that urban areas not only provide an opportunity for encounters with people different from us, but also open up the possibility of remaking ourselves in ways that may not be possible in homes or traditional communities. Reproduced courtesy of Woody Wood.

the concentration of LGBTQ residents and businesses catering to them in specific urban neighborhoods. Well-known examples of these include Boy's Town in Chicago, the Castro District in San Francisco, Toronto's Village, and Vancouver's Davie Street Village. These neighborhoods have historically provided a space for the production of identity and community, not only for gay men but also (albeit to a lesser extent) for lesbian women and gender-queer people. Gay enclaves also serve as a visible marker of the existence and strength of the LGBTQ community to the larger public. As Catherine Nash and Andrew Gorman-Murray (2014) explain, these neighborhoods provided a venue in which 'gay movement politics promoted support of gay businesses as a political act in tandem with arguments around the need to protect and preserve political territory for economic and social strength' (759). Increasingly, queer identities are also being accepted outside of these enclaves, as expanding spaces of LGBTQ residence, leisure, and business are gradually blurring the boundaries of gay/straight space (Browne and Bakshi 2011).

While urban public spaces have been theorized as having unique possibilities for the expression of marginalized identities, some research suggests that privatized public spaces (such as malls, coffee shops, etc.) can be equally, if not more, emancipating, depending on the geographic context. Nazgol Bagheri (2014), for example, discusses affluent Iranian women's use of privatized public spaces, arguing that 'women in Tehran use modern and often privatized public spaces such as shopping malls in unexpected ways to enjoy a sense of freedom that is not usually available in traditional urban locations' due to heightened public scrutiny of them as women in public (1298). And indeed, in this context, women's presence in these types of spaces also shapes their meaning: their mixed-gender atmosphere indicates their modern, global, and cosmopolitan nature (de Koning 2009). However, while affluent women enjoy freedom in such spaces, malls and other privatized public spaces often remain off-limits to many women, including migrant workers, the poor, and ethnic minorities. Indeed, it is often the precise absence of people from other class, ethnic, or racial groups that results in these spaces being considered safe and appropriate for affluent women's use. As discussed in Chapter 3, similar dynamics play out in many urban locations in Latin America and Africa, where commercialized public spaces such as shopping malls are highly segregated by class and, as such, are seen as safe spaces for (affluent) women and children to recreate while pure public spaces such as parks and streets are considered risky. This example highlights both the possibilities of urban spaces, as well as their potential to reinforce and re-inscribe gender, sexual, and class norms.

Conclusion

This chapter highlights the ways in which urban areas are sites of critical—and contradictory—gendered processes. Far from viewing cities as wholly liberating or exclusionary for women, contemporary urban feminist geography recognizes the complex ways in which gendered experiences and identities arise from, and help structure, urban processes. In particular, feminist geography helps us to understand the connections among individual identity, social experience, and processes occurring at the urban scale, such as the creation and use of public space and neoliberal urban policy. Women's intersectional identities are critical to how they engage and experience the city: experiencing the city as a safe place for living, working, and recreating often reflects a privileged race and class position. These privileges are supported through urban policy and the creation of privatized or gentrified spaces reflecting the interests of the elite.

Nonetheless, feminists and other activists continue to confront, resist, and change gendered exclusions in urban areas. In addition to activism surrounding women's access to public space, women are also at the forefront of collective attempts to challenge the marginalizing effects of neoliberal urban policies for themselves and their communities, through protest, community building, and arts intervention. The challenges of modern urban space are being confronted as well through individual actions in people's everyday lives, including walking confidently, enjoying public leisure spaces in new ways, and surviving neoliberal urban change day by day. Through these collective and individual actions, those traditionally left on the margins of city spaces insist on their right to participate in, and appropriate, urban space.

Recommended reading

Bondi, Liz, and Damaris Rose. 2003. 'Constructing Gender, Constructing the Urban: A Review of Anglo-American Feminist Urban Geography.' *Gender, Place & Culture* 10 (3): 229–45. doi:10.1080/0966369032000114000.

Chant, Sylvia, and Cathy McIlwaine. 2015. *Cities, Slums and Gender in the Global South: Towards a Feminised Urban Future*. New York: Routledge.

Day, Kristen. 2006. 'Being Feared: Masculinity and Race in Public Space.' *Environment and Planning A* 38 (3): 569–86. doi:10.1068/a37221.

England, Marcia R., and Stephanie Simon. 2010. 'Scary Cities: Urban Geographies of Fear, Difference and Belonging.' *Social & Cultural Geography* 11 (3): 201–7. doi:10.1080/14649361003650722.

Fincher, Ruth, and Jane Margaret Jacobs. 1998. *Cities of Difference*. New York: Guilford Press.

Huang, Shirlena, and Brenda S. A. Yeoh. 2008. 'Heterosexualities and the Global(ising) City in Asia: Introduction.' *Asian Studies Review* 32 (1): 1–6. doi:10.1080/10357820701871187.

Nash, Catherine J., and Andrew Gorman-Murray. 2014. 'LGBT Neighbourhoods and "New Mobilities": Towards Understanding Transformations in Sexual and Gendered Urban Landscapes.' *International Journal of Urban and Regional Research* 38 (3): 756–72. doi:10.1111/1468-2427.12104.

Panelli, Ruth, Anna Kraack, and Jo Little. 2005. 'Claiming Space and Community: Rural Women's Strategies for Living With, and Beyond, Fear.' *Geoforum* 36 (4): 495–508. doi:10.1016/j.geoforum.2004.08.002.

Peake, Linda, and Martina Rieker, eds. 2013. *Rethinking Feminist Interventions into the Urban*. New York: Routledge.

Phadke, Shilpa, Sameera Khan, and Shilpa Ranade. 2011. *Why Loiter? Women and Risk on Mumbai Streets*. New Delhi: Penguin Books India.

Secor, Anna. 2004. '"There Is an Istanbul that Belongs to Me": Citizenship, Space, and Identity in the City.' *Annals of the Association of American Geographers* 94 (2): 352–68.

Spain, Daphne. 2014. 'Gender and Urban Space.' *Annual Review of Sociology* 40: 581–598. doi:10.1146/annurev-soc-071913-043446.

Chapter 5

GENDERED WORK AND ECONOMIC LIVELIHOODS

Ann M. Oberhauser

Chapter 5

Oberhauser

Introduction

Charlotte and her family live in a small town in rural West Virginia. She works two minimum-wage, part-time jobs, one at a local Walmart and the other in a retirement facility for the elderly. In order to pay for increasing household expenses, she recently started to work at the retail store. In addition to these jobs, Charlotte and her mother share childcare responsibilities for her two young children. Public transportation is not readily available in this area, so she uses her mother's car to drive to work.

Charlotte's husband is an independent truck driver who hauls lumber for timber companies in the state and is gone at least three days a week. The family doesn't have benefits or health insurance through their employers, so they rely on government support for their healthcare needs.

(Field notes, May 21, 2004)

This family's efforts to balance multiple economic activities, childcare responsibilities, and transportation illustrate how **work** is connected to social relations in very fluid and contested ways. Charlotte's family uses diverse strategies to generate income and to negotiate social relations that are embedded in gender, class, generational, and other social identities. Intra-household and familial dynamics are key to understanding how this family, and many others, juggle care work and other aspects of **social reproduction**. Furthermore, the precarious nature of work in this family is connected to the multiple and shifting economic opportunities that have resulted from economic restructuring and changing labor relations, which are partly due to the decline of jobs in the coal fields of West Virginia.

From a feminist perspective, these and other forms of economic strategies confirm that production is embedded in the social reproduction and labor of families, communities, regions, and even global economies. This chapter examines the negotiation of diverse **livelihoods** or **economic strategies**, such as those highlighted in the scenario from West Virginia, through a feminist geographic lens. Livelihoods, as defined by Elizabeth Francis (2000), represent the ways in which household members 'make a living' through a complex set of activities and social interactions (76). The discussion draws from this concept of economic livelihoods to explore who does what kind of work, and how these activities are linked to socioeconomic and historical patterns in society. The integration of these practices and social relations, beyond and across geographic boundaries, is presented in this chapter as an important dimension of gender and economic strategies.

Gendered work and economic livelihoods

Feminist analyses of work and the economy include attention to the multiple ways in which people support themselves by providing food, shelter, and other daily needs that contribute to the well-being of individuals, families, and communities. These strategies are in turn grounded in geographies of power and social inequalities that affect how people participate in economic livelihoods. Linda McDowell (2013), Susan Hanson and Geraldine Pratt (1995), Katharyne Mitchell, Sallie Marston, and Cindi Katz (2004), and other feminist economic geographers have analyzed the gendered nature of social reproduction and its connection to **gender roles** and **divisions of labor** in these diverse economic strategies. In other words, reproductive activities and social relations influence the type of labor that men and women conduct to generate income and support themselves and their families. For instance, women are disproportionately found in occupations that involve **care work**, such as nursing or childcare, because of their association with reproductive labor in the household. As noted by Katie Meehan and Kendra Strauss (2015), labor of all forms is connected to social reproduction in what they describe as 'the interaction of paid labor and unpaid work in the reproduction of bodies, households, communities, societies, and environments' (1).

This chapter focuses on different aspects of feminist geography's approach to the economy, work, and social reproduction. The discussion examines the diverse ways in which people gain control over their life's work, through examples from the Global South and North. The second section introduces how gender and work are conceptualized in feminist geography, with attention to production and social reproduction, intersectionality, informal work, and globalization. This discussion is followed by analyses of the gendered labor process and specifically how labor is valued through gender differences in occupations, wages, and the status of work. The fourth section addresses intersections of globalization and gender with an emphasis on work, inequality, and migration. The concluding section outlines strategies for change and resistance to marginalization through transgressive movements at local, national, and global scales.

Conceptualizing gender and work

As noted above, feminist geographers examine economic strategies and the power relations that impact, and are reflected in, social reproduction and production. Conventional approaches to the economy tend to ignore how social identity shapes wages, types of employment, and other work-related issues.

Furthermore, these mainstream economic analyses often neglect the rich geographic and historical contexts that lay the foundation for the economic marginalization of women and the racialization of the labor force, for instance. In many ways, feminism rewrites our understanding of the economy with attention to links between individuals, households, and more macro scales of production and labor. Feminist analyses of the economy also carry a political and ethical commitment to creating critical responses to the inequalities that stem from neoliberal economic theory and practice.

The discussion starts with a somewhat familiar historical account of how capitalism laid the foundation for contemporary social relations in the Global North, and particularly in Europe. With the advent of the Industrial Revolution in the late 1700s and the spread of industrialization and urbanization in many parts of Europe, economic activities became somewhat segregated as waged labor took place in the productive arena, and unpaid labor occurred in the domestic realm. As noted in Chapter 3, in this context, men were more likely to enter the paid labor force while many women remained in the domestic sphere. The gender divisions of labor highlighted during this transformative period in Europe were acutely impacted by other social and economic identities. Feminist geographer Doreen Massey (1995) examines how class and gender intersected with certain economic practices during the early industrial era of the UK. She argues that working-class women and women of color were more likely to be employed in factories than wealthier women and white women.

This analysis is related to the connection of **productive labor** (income generation in the public sphere) and **reproductive labor** (unpaid reproductive activities in the domestic or private sphere). The binaries that are often associated with productive and reproductive work have been challenged through recent feminist work on social reproduction, or what Mitchell et al. (2004) describe as 'life's work.' According to Marxist political economy, reproductive labor (often women's labor) is instrumental in sustaining the waged workforce. This caretaking, or the provision of food and other basic needs, supports the capitalist system.

Another thread of feminist economic geography examines the continuum of informal activities, ranging from subsistence agriculture, to petty trade in street markets, and the unregulated provision of services such as childcare. In many areas, women are disproportionately involved in **informal-sector activities** due to marginalization from formal-sector jobs, a lack of training, or an inability to access credit (Box 5.1). Ruth Pearson (2007) examines the

Gendered work and economic livelihoods

Box 5.1 Women in the informal sector

The informal sector is crucial to the economy of households and communities in many parts of the Global South. These activities are often gendered in terms of who does this work, and how much income is generated for both individuals and families. In rural South Africa, for instance, it is not uncommon for women of African descent to help support their families by engaging in informal economic activities (Oberhauser 2016). The woman in this photo is selling vegetables and other produce at a local roadside market while her husband works in the city of Johannesburg as a migrant laborer. Remittances from his earnings provide some resources for their household, but are not enough to fully support their expenses. Earnings from her informal activities are critical in paying for school fees for their children, and purchasing clothing and food for the family.

Figure 5.1 A woman selling farm produce in Limpopo Province, South Africa, 2012. Reproduced courtesy of Ann Oberhauser.

Activity

Consider these discussion questions to reflect on informal activities where you live:

1. In your community, what types of informal activities do people engage in as a means to earn some income? (For instance, are people involved in babysitting, lawn mowing, supplying fire wood, or the illegal sale of drugs?)
2. Are there gender differences in who does what type of work?
3. Consider the monetary value placed on these activities and how this is gendered.

prevalence of women working in the informal sector throughout the world. According to her analysis, 'the bulk of the female labour force in the majority world do not labour within a formal economy regulated by national laws concerning pay, working hours, leave, or indeed have any access to work-related social protection' (202). See Chapter 4 for further analyses of the role of the informal economy and labor in urban economies.

Feminist economic geography also challenges assumptions about who does what kinds of work and how this varies among different societies. Groundbreaking research in this field by J. K. Gibson-Graham (2006) recognizes important alternatives to the capitalist system of valuing and performing labor that are instrumental in sustaining people. These diverse economies involve the provision of basic needs, as well as income generation for people who do not participate fully in the capitalist economy, in ways that they refer to as 'remaking the economy in alternative terms' (xi). Through this and similar frameworks, feminism expands our understanding of the economy in strategies that include barter exchange of goods and services, subsistence livelihoods, and informal-sector activities. These activities often entail small-scale and low-productivity economic activities that are not subject to state regulation or taxation.

Intersectionality also provides insight into how diverse social categories or identities collectively impact the different ways in which people participate in both social reproduction and the paid labor force. Using an intersectional framework, gender divisions of labor and power relations are integrated and co-produced with other social identities, such as race, age, ethnicity, sexuality, and class. As noted in Chapters 2 and 3, in discussions of the body and culture respectively, these multiple and often intersecting social identities form the basis for understanding access to work and the overall status of workers. One of the ways of understanding these intersections is through analyses of the value that capital places on human difference. As argued by Ruth

Wilson Gilmore (2007), Laura Pulido (2016), and other scholars in critical ethnic studies, race and ethnicity are the basis for some of these differences. This framework considers 'dominant historical narratives of racism' (Pulido 2016, 4) that assign differential values placed on workers with diverse racial and ethnic identities. Empirical studies of paid labor show, for example, that women of color tend to have a larger wage gap than women as a whole. Hispanic women, or Latinas, earned nearly 55 percent and African American women earned approximately 60 percent of the median annual earnings for white men in 2014. These earnings compare to a 78 percent wage gap for all women (Hegewisch and Hartmann 2015). Thus, intersecting social identities such as race and gender influence people's economic activities. Legacies of discrimination and marginalization in certain types of employment help to explain and historically situate material conditions, including the wage gap among women, and between women and men.

Another important aspect of feminist economic geography is the emphasis on unequal spatial and social dimensions of how women of diverse backgrounds engage in economic activities. This discussion underscores how **uneven development** or the spatial inequality produced by capitalism in various contexts relates to racial, gender, and class-based differences. In Nairobi, Kenya, for instance, it is not unusual for wealthy urban neighborhoods to be located near impoverished slums. Efforts to privatize public services and reduce government expenditures are indicative of a neoliberal economic restructuring that accentuates this spatial inequality. As a result, slum dwellers in Nairobi face difficulties accessing public services—such as sanitation and transportation—that are readily available in nearby wealthy areas (Njeru et al. 2014). Furthermore, women in low-income households are often responsible for social reproduction, such as cooking, cleaning, and childcare, and thus are disproportionately impacted by this lack of access to public services. This example demonstrates that inequalities and uneven development concerning work and gendered social relations are linked to the broader economic conditions and contradictions of neoliberal capitalism.

Finally, spatial differences in the cost and skill levels of labor are closely related to globalization or shifting economic activities among different countries. Economic outsourcing often entails the transfer of production and services to countries and regions where labor is cheaper, less regulated, or non-unionized (Beneria et al. 2016). Companies take advantage of these differences in labor costs, which, in turn, enhances the global economic restructuring of production. Women and other marginalized workers are often employed in these activities because they are the lowest-paid and most vulnerable workers in many parts of the Global North and South. The expansion of capital, discussed in more detail below, reflects these dynamic situations, further contributing to social, economic, and spatial disparities.

Valuing gender and other social identities in the labor process

Society places a certain value on work, or the production of goods and services, that is closely tied to social identities of workers (such as their gender, race, ethnicity, and age). The compensation that people receive for doing a job, as well as the status or prestige associated with that job, are often grounded in specific geographic contexts and assumptions about who does what and the values that are placed on these skills and activities. Numerous feminist geographers have examined how gender and other aspects of social identity impact work (Oberhauser 2000). For example, McDowell's (2003) research on young working-class white males explores their shifting identities in the economic landscape of post-industrial society in the UK. In particular, her work demonstrates how their masculine identities are challenged as they experience increasing difficulties finding employment, with the loss of manufacturing and other male-dominated jobs. In some of McDowell's earlier research on gender and the economy, she examines gender stereotyping and the influence of particular class and gender attributes in London's merchant banking sector. She describes the gendered performances of men and women in the workplace as 'inscribing gendered characteristics on the body in ways which conform to or transgress expected patterns of behavior in a particular cultural milieu and physical settings' (McDowell 1997, 133). Much of this gendered performance is displayed in clothing, hairstyles, make-up, and language, and sets up a double standard for women and men. According to McDowell, women are often socialized to be more conscious of their appearance in the workplace than men. This work has been modified in recent years to examine how gender performance in the workplace can also negatively affect men.

As noted in Chapter 2, Judith Butler's (1993) research on social identities in the work sphere and elsewhere also shows how bodies are disciplined and regulated into normative performances. Heteronormative workplaces, for example, construct other sexual identities and gender performances as problematic and unacceptable. Consider, for instance, typical workplace discussions about weekend plans with family, pictures of loved ones, or bringing spouses to the company picnic. Gill Valentine (1993) and, more recently, Petra Doan (2010) discuss how the perceptions and experiences of everyday spaces may serve to discipline workers into dominant heteronormative relations while marginalizing other forms of relationships that transgress these gender norms. Another example of cisgender or heterosexual dominance in the workplace is policies concerning partner benefits that may not extend to same-sex/transgender partners. These and other biases in the social relations of work underscore the prevalence of institutionalized discrimination.

Historically, gender-segregated roles in society have often changed with the introduction of technology in the production process and service sector. In the US, clerical and secretarial jobs were dominated by men in the 18th and 19th centuries, but have become dominated by women since the early 20th century. Feminist geographers Kim England and Kate Boyer (2009) examine gender shifts in the financial services sector over the past century as they relate to changing technology and information. These researchers argue that the growing need for low-paid, routinized office jobs in this sector coincides with companies hiring more young and single women. Thus, the financial services sector is an example of the new geographies of both job mobility and stagnation that differ by sex and by scale. Gender roles in the workforce are also culturally and historically embedded in shifting patterns of work and technology.

The dynamic landscapes for employment in the financial services sector relate to research in feminist economics and geography on the growing incidence of **precarious labor**. The new neoliberal labor force is one where uncertainty in terms of continuity, income, and protection create vulnerable work environments for labor (Pollard 2013). Precarity is mostly experienced by already marginalized workers who are limited to temporary and contingent work, experience deskilling of their labor, and are most vulnerable to layoffs and redundancy (Meehan and Strauss 2015).

Some of the vulnerabilities and disparities among workers in the labor force reside in the production of social difference under capitalism. The economy operates in a way that both creates and builds on difference among, and within, groups of labor. As noted above, critical ethnic studies have developed important work on the relationship between racism and capitalism, or racial capitalism. Pulido (2016), for instance, highlights the racial dimensions of capitalism in ways that 'place contemporary forms of racial inequality in a materialist, ideological and historical framework' (4). Thus, racial capitalism draws from feminist work to analyze the process of dehumanizing non-white bodies under capitalism. See Chapter 7 for further discussion of the links between racial capitalism and environmental justice, or disparities in the impact of pollution and environmental destruction.

This chapter argues that gender and other social categories impact, and are influenced by, the performance of and value placed on labor. Previous chapters demonstrate how men and women have traditionally assumed certain gendered tasks or roles in the home. These roles, in turn, carry over into the workplace and impact unequal power relations that are often reflected in wages and salaries. Diane Perrons (2004) has written extensively about growing inequality within the 'new economy' of the 21st century. According to her research, women and people of color are more apt to work in part-time jobs, service-sector occupations, and overall low-paid areas of the labor force. The following discussion emphasizes two areas where gender differences are evident in the workplace: **occupational segregation** and the **wage gap**.

In many societies, women and men tend to be segregated or concentrated in certain occupations that are linked to societal norms concerning gender roles and identities (Hanson and Pratt 1991). Predominantly female occupations are those in which women represent at least 75 percent of workers and predominantly male occupations are those in which women hold fewer than 25 percent of the jobs (Hegewisch and Hartmann 2015). As illustrated in Figure 5.2, women as a whole are disproportionately represented in social services and care-giving jobs, such as elementary and middle-school teachers, social workers, and speech-language pathologists. In contrast, they are underrepresented in high-level management (e.g., chief executives) and professional positions (e.g., physicians, surgeons,

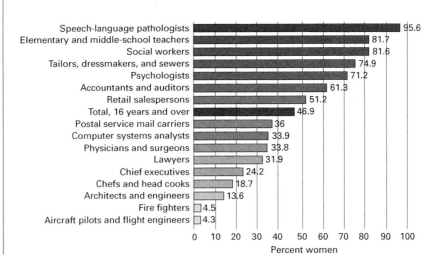

Figure 5.2 Women as a percentage of total employed in selected occupations, 2011. Source: US Bureau of Labor Statistics (www.bls.gov/opub/ted/2012/ted_20120501.htm)

Gendered work and economic livelihoods

lawyers, and fire fighters). The persistent segregation of women since the early 1970s in certain occupations tends to reflect how society constructs feminized forms of labor that include caregiving (teachers and speech-language pathologists) and service work (dressmakers and clerks).

Gender segregation in the workforce and among certain occupations continues to have significant implications for women. Studies demonstrate that women are less likely to enter a variety of higher-status occupations, especially in the science, technology, engineering, and mathematics (STEM) fields (Hill et al. 2010). Parallel situations can be seen in education, where, for example, girls are often underrepresented in the sciences or mathematics in primary and secondary schools. This is not due to the fact that girls are less capable than boys in math or science, but because they tend to have less confidence in their science abilities than boys. (See studies by the US National Academy of Sciences (2007) and Ceci et al. (2014) for more details about these patterns of gender bias in education and careers.)

Analyses of gender and work also challenge dominant norms about men's and women's positions in traditional fields. For instance, Figure 5.3 is an advertisement in which images of male nurses confront certain stereotypes

Figure 5.3 Occupational stereotypes: the case of nursing. Reproduction courtesy of the Oregon Center for Nursing (oregoncenterfornursing.org).

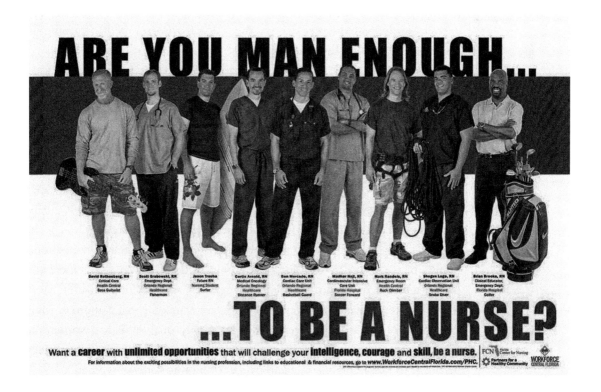

about masculinity and femininity. In this case, men are employed in a female-dominated job. The question 'Are you man enough . . . to be a nurse?' is raised in the tag line about nursing as a field with 'unlimited opportunities that will challenge your intelligence, courage and skill.' The men in the image reinforce masculine stereotypes that include physical strength and participation in such activities as golf, guitar playing, and rock climbing. Images like this challenge the stereotyping of nursing as a feminine occupation, as well as transgressing ideas about masculine roles.

In addition to the gender differences in the types of jobs that men and women hold, research on the wage gap shows that pay inequities between women and men and among women of different racial and ethnic backgrounds persist in many fields and continue to impact women's marginalization in the workplace. Feminism, and feminist geography in particular, analyze the sociospatial dimensions of the wage gap. As illustrated in the opening scenario to this chapter, some women in rural areas have more difficulties finding adequate work and face lower earnings than women in the US as a whole (Hegewisch and Hartmann 2015; Oberhauser 2002). Women in West Virginia, for example, have a larger wage gap (69 percent) than women in the US as a whole (79 percent) (Oberhauser 2000). Thus, women in rural states, especially those in the southern US, tend to have lower wages and a greater **gender earnings ratio**—or the ratio of women's median earnings to men's—than women living in urban areas in the northern US.

These geographic disparities in wages are impacted by gender as well as race, ethnicity, age, and other social categories that reflect societal norms and perceptions of work and how society places different value on labor. Women tend to have lower-paid jobs, often without job security or support, in part because their labor is not valued as much as men's. These wage gaps are, in turn, linked to the concentration of women in minimum-wage jobs. Recent debates in the US about the need to raise the minimum wage highlight the importance of this measure to lift women and children out of poverty. (See the organizing that is done around the need to increase the minimum wage such as the Fight for 15 movement (www.fightfor15.org) campaigning to raise the national minimum wage to $15 an hour.) According to the Children's Defense Fund (2014), a federal minimum-wage job paying $7.25 an hour for a full-time, year-round worker with two children provides a family income that is approximately two-thirds of the federal poverty level for family of four. Since women are disproportionately employed in minimum-wage jobs, this debate is especially pertinent to raising the income and overall economic status of women.

Gendered work and economic livelihoods

In sum, this discussion shows that economic discrimination, social inequality, and uneven spatial patterns are evident in the work experiences of both men and women. Women, and particularly women of color, tend to experience the greatest inequalities in the workplace. Conditions vary, however, by location, historical period, and social context in ways that impact occupations, wages, and other opportunities for advancement among women. The next section examines globalization as a process where labor and work are impacted by increasing integration and inequality in the economic sphere.

Gender, work, and globalization

'Since the late 1970s, the process of global economic integration has become a most powerful source of change driving national economies and affecting all aspects of social, political, and cultural life' (Beneria et al. 2016, 93). Indeed, one needs only to observe the widespread production and consumption of technology manufactured by global corporations such as Apple, Samsung, and Nokia to appreciate the economic and cultural impact of **globalization**. The ubiquitous nature of mobile phones in many developing countries is linked to changing consumer habits, communication, and economic opportunities. This example highlights differing views about the role of globalization in providing economic opportunities and shaping social and cultural interactions. Some experts argue that globalization has a positive effect on the economy in terms of growth and rising incomes. Others argue that it has negative consequences associated with rising social and spatial inequality. This section focuses on the relationship between globalization and gender as a way to understand how this growing integration of cultural, political, and economic activities is measured, experienced, and challenged.

The following exercise (Box 5.2) situates this discussion in the context of women and economic empowerment. A growing number of international organizations and programs focus on issues that will empower women and improve their economic status, and therefore their overall status, in different societies. In this exercise, the organization UN Women provides data about the status of women throughout the world and identifies ways to challenge these barriers to gendered economic parity.

Different approaches in feminist geography to development and economic crises examine how **neoliberal globalization** shapes the material and ideological forms of economic activities. As noted in Chapter 4 on neoliberalism in an urban context, neoliberal globalization is the implementation of policies

Box 5.2 Women's economic empowerment

Many global organizations, such as the United Nations Entity for Gender Equality and the Empowerment of Women (UN Women), address goals that support work based on equal pay, non-discriminatory conditions, and shared access to land and other basic resources.

(For a full description of this program, see www.unwomen.org/en/about-us/about-un-women.)

Activity

Examine the information about women and economic empowerment in the United Nations document, on the website below. Read through each section and answer the questions in this exercise.

> www.unwomen.org/en/what-we-do/economic-empowerment/facts-and-figures#notes

1. Name at least three of the main challenges or barriers to women's economic advancement as outlined in this report. How do these barriers vary by region? Give some examples of these regional differences. In your opinion, what are some of the explanations for these geographic differences in the economic status of women?
2. According to this report, what measures or steps can be taken to improve women's economic status? Think about the scale of these changes as they take place at the household, local or community, national, or global levels.
3. How do countries and societies as a whole benefit from the economic advancement of women? What adjustments or changes in gender roles are needed to bring about some of these advancements?
4. Can you relate to some of the themes and measures discussed in this report? Have you, or has someone you know, experienced some of the occupational segregation, discrimination, or challenges discussed here? Explain.

Gendered work and economic livelihoods

that encourage increased privatization of capital, reduce government and public spending, and contribute to global financial crises, such as those in 1998 and again in 2008. Jane Pollard (2013) analyzes these economic and financial crises through the lens of feminist economic geography. She explains how the crisis that resulted in the severe economic impacts of the Great Recession of 2008 was linked to deregulation and subprime lending practices in developed countries. The outcomes of this economic recession devastated many working- and middle-class households, whose debt increased significantly, leading to a crisis in consumer demand, investment, and eventually labor. Although men (especially white men) in the finance sector lost jobs in higher proportion than women, people across the country lost houses, and savings and retirements were disrupted.

At the global scale, the International Monetary Fund and other international institutions, such as the World Bank and the International Labor Organization, tend to support neoliberal measures so countries will lower expenditures on social services, and thus reduce their national debts. These initiatives are widely known as structural adjustment programs (SAPs). Victoria Lawson (2007) discusses how these measures have negatively impacted economic growth. She specifically argues that neoliberal globalization and accompanying policies have further marginalized women, who often have primary responsibility for social reproduction of households and communities. Their earnings and expenditures, for instance, are more likely to pay for the education of children and family health care.

Women's work in global contexts

One common characteristic of globalization is the substantial fragmentation or separation of different parts of the labor process. This fragmentation drives capital to locate in areas with low labor costs, minimal protection and benefits for workers, and cheaper inputs to production as a whole. As discussed above, outsourcing also leads to shifts in the production of goods and services to countries in the Global South. Women are among the most vulnerable and lowest-paid workers in many of these regions, where patriarchal norms devalue women's labor and their overall status in society.

The underlying dynamic of global capital and labor processes depends on changes that affect production and communication in both host and home countries. For instance, the globalized, labor-intensive form of manufacturing garments, footwear, and electronics has developed alongside the growth of Export Processing Zones (EPZs), which are zones set up by governments in developing countries in order to manufacture and export industrial goods, as well as developing activities in the service sector. Local governments in these areas often provide incentives and subsidies to encourage foreign investment in manufacturing and low-skilled service sectors employing cheap labor (Figure 5.4). In Southeast Asia, Ramkishen S. Rajan and Sadhana Srivastava (2007) have conducted studies about women's employment in these EPZs and industrial parks. United Airlines and Dell Computers are two of the largest employers of women from local and surrounding rural areas in call centers, customer service, and data entry positions in Malaysia. This offshore and outsourced economic activity often employs a precarious labor force with few regulations and highly exploitative working conditions. Young women employed in these sectors also often face the social stigma of living in dormitories away from their families and social support.

Melissa Wright's (2006) research on the *maquiladora* industry in the border region of the US and Mexico draws attention to global industrialization in this region and its impact on the working poor (especially women). She refers to the myth of the 'disposable Third World woman' as someone who, over time, loses her ability to work and is worthless, discarded, and replaced by the factory that employs her. Again, this is an example of the precarious nature of labor and how social identity is tied up with workers who are overlooked in

Figure 5.4 Workers at the Mustang socks factory in Palghar, India. Photograph by Atul Loke. Reproduced courtesy of Panos Pictures. (www.panos.co.uk/preview/0022 6049.html?p=1)

the globalization of capital. Thus, women's labor in this context is valued by industry only to the extent that it contributes to profit. The women themselves are essentially 'disposable' to global corporations as they look for more cost-effective sites of production and cheaper labor.

Furthermore, some companies with global production facilities frequently choose to locate themselves in countries where English is widely spoken. These capital investments tend to reinforce **cultural hegemony** and sometimes renegotiate Western identities in global labor. In *High Tech and High Heels in the Global Economy*, Carla Freeman (2000) emphasizes this social aspect of transnational capitalism in her study of Afro Caribbean women working in high-tech informatics jobs in Barbados. Freeman's work demonstrates how these workers redefine companies' approaches to the 'ideal' employee and create unique 'pink-collar' identities.

Similar research by Reena Patel (2010) examines Indian call centers in the business-outsourcing industry, where masculine and feminine identities are incorporated into the labor process. Specifically, cultural considerations such as language, accent, and other social practices become integrated with the employees' abilities to answer calls from customers in the Global North.

Finally, globalization impacts local gender relations, in particular the economic status of households and communities through investment in manufacturing and other types of economic activities. Some feminists argue that this investment in developing countries entails highly exploitative labor conditions for female workers, who are often separated from their homes and families (Beneria et al. 2016). In contrast, Naila Kabeer and Simeen Mahmud (2004) maintain that Bangladeshi women in the industrial sector, for instance, gain independence and play a larger role in their household decision making with this added income. Money earned from these jobs contributes to households and provides support for education, housing, and other material and social needs of their families. These cases demonstrate how globalization may positively impact household gender relations, as women gain status and are able to make decisions about expenses. Whatever the benefits or drawbacks, the incorporation of women into the global economy has far-reaching impacts on their lives and those of the people in their communities.

Gendered mobility of labor: global migration

The growing mobility of workers is also linked to the increasing globalization of capital. People migrate from their countries of origin in order to access economic, as well as social and political, opportunities. In many cases, migration is a highly gendered and racialized process, in both the type of work people conduct, and in the processes driving them to leave their homes. Feminist geographer Sylvia Chant (1998) distinguishes between permanent and temporary migration made up of movements of labor from the Global South to the Global North, as well as from rural to urban areas. Migrant labor is often employed in host societies based on certain constructions of gender, race, and other social identities. For example, domestic workers, maids, and service providers tend to be female migrants, while male migrants are often employed in the fields of construction, landscaping, and sanitation (Beneria et al. 2016). These trends are partly a result of the high proportion of women who work outside the home in many host countries, and the need to hire (or the privilege of being able to hire) people to take care of their children. Parents who face increasingly expensive childcare costs in areas with decreasing state support due to neoliberal policies often turn to service providers from the Global South.

Much of this female migrant labor in domestic and other service jobs is part of what Geraldine Pratt (2012) and Kim England and Caitlin Henry (2013) refer to as **care work**. Global care commodity chains, such as nursing and domestic labor, involve migrant women who care for children and provide domestic labor in wealthy households of the Global North. In her book *Families Apart*, Pratt (2012) offers an in-depth look at migrant mothers who experience personal conflict between their roles as domestic laborers in Vancouver, Canada, and the families and children they leave behind in the Philippines. In these cases, racialization of the labor market is equally important to the gendered aspects of care work among migrant laborers. Stories of harassment and abuse by families and employers in these host countries are common. Likewise, the migration of nurses to the Global North has implications for both sending and receiving nations in this care work across borders (England and Henry 2013). Workers often face low pay, the loss of basic rights, and difficulties traveling home. These hardships often lead to tremendous isolation, as women are away from their own families for months and even years at a time.

Global trafficking of humans has also become an important focus for feminist activists and researchers who explore this aspect of labor mobility. Work by Vidyamali Samarasinghe (2012) shows that people who experience this kind of involuntary bondage are often misled into believing that they are moving to decent jobs. These situations have reached all-time highs, with demand for exploited labor or sex workers increasing in both the Global

Gendered work and economic livelihoods

North and South. As the most vulnerable members of society, women of certain racial, class, and ethnic backgrounds find themselves among those forced to work as prostitutes or laborers in areas with minimal or no pay. The lack of legal status among these women increases their vulnerability at the hands of traffickers and corrupt law enforcement personnel. Traffickers realize enormous profits from this growing form of labor exploitation. A Eurostat study found that global profits from sex work reached $27.8 million during the 1995–2004 period, and the number of trafficked persons increased 18 percent between 2008 and 2010 (Belser 2005). As a response to the heightened awareness of these circumstances, legislation has been introduced in the US to make trafficking of humans illegal in many states.

Within this analysis of migrant workers lie important debates about their motivations and forced or voluntary decisions to leave their home countries. As discussed in Chapter 6 on geopolitics, these motivations are highly gendered and racialized. Groundbreaking work by Alison Mountz (2004) examines the contested and fluid status of human migration and, particularly, smuggling. Her work challenges conventional approaches to discourses around the detention of migrants by the nation-state, and instead emphasizes the role of corporeal geographies that focus on the scale of the body as a site of contested power relations. Finally, remittances are important aspects of labor migration, and contribute significantly to local economies and households. Thus, social and economic geographies are crucial to understanding many migrant laborers' decisions and experiences.

Global resistance and strategies for change

This section examines globalization through a feminist lens, with attention to the social inequalities and spatial disparities that are often the focus of narratives about gender and globalization. Critical feminist analyses of globalization provide the platform to challenge these obstacles in numerous contexts and at multiple scales. The discussion emphasizes transgressive strategies that engage women and communities in developing progressive and positive change. Indeed, the mobilization of feminist voices can support increased resistance and protest against the exploitation that accompanies neoliberal global economic processes. In the examples highlighted below, micro-entrepreneurs, factory workers, and farm laborers enact change by leveraging resources and resisting marginalization. In some cases, women who are part of the global economy are better situated than men to transform households and communities.

The analysis focuses on three strategies of resistance that have the potential to create change in the face of growing economic globalization: networks and collaborative alliances; local ownership of production; and sustainable models of trade and production of goods and services.

First, economic and social networks connecting individuals and communities at local and global scales provide important means of creating and coordinating efforts to support women. One example is **microfinance**, whereby governmental and non-governmental organizations provide small loans to lower-income and impoverished people for investment in economic activities. Women form the majority of small-scale borrowers because they often do not have access to capital or other means of leveraging loans. The intentions and outcomes of microfinance have been highly contested in gender and development literature. Some geographers argue that microfinance often drives women further into debt with high interest rates (Young 2010). Others outline the advantages to women of being able to access credit, and thus empower themselves in their families and communities. For them, these networking strategies are a means of cultivating transgressive practices among women in microfinance and microenterprises. Their research proposes alternative approaches that 'engage women's agency in progressive ways' (Shakya and Rankin 2008, 1228) and reflect widening opportunities for women.

Networking among individuals and communities also expands opportunities for people to connect with others outside their region. For instance, the Self Employed Women's Association (SEWA) in India is a network that provides social training, housing, and banking to its members. These broad areas of support are invaluable to many women living in difficult circumstances without adequate assistance from family or government. Sabhlok (2010) has conducted in-depth research on the use of scale by SEWA to represent and effect change at the local level, while becoming a powerful voice at both the national and the global levels in advocating for working-class women in India. Another program that aims to economically empower women and other impoverished communities is Conditional Cash Transfer (CCT) in Brazil and other countries in Latin America (de Janvry and Sadoulet 2006). These funds, donated by governments and donor organizations, help families in times of extreme scarcity and economic crisis. This system of cash transfers requires that the money generated from these funds is invested in children's human capital, such as education and health. The effectiveness of this model inevitably varies by the level of literacy of the parents, the age and gender of children, and other factors within these households.

Gendered work and economic livelihoods

Second, globalization coincides with growing efforts among local producers to combat capitalist exploitation and take control of their labor through unionization and ownership of production in agriculture, manufacturing, and services. Alta Gracia Apparel in the Dominican Republic, advertised as a 'living wage company,' is an example of workers' resistance to exploitation by big corporations. They claim to be one of the only apparel companies in the Global South that is certified as a business that pays their workers enough to meet the basic needs of their families. The promotional material of Alta Gracia Apparel states they are 'paying workers enough so that they can provide their families with food, clean water, clothing, housing, energy, transportation, child care, education, and health care' (Alta Gracia Fair Trade 2016). These efforts are embedded in gendered relations, partly because women make up the majority of workers in the global garment industry.

Angela Hale and Jane Wills' (2007) multi-year study of women in the global garment industry highlights how women work in poor conditions with low pay, forced overtime, and long hours. Many critiques of this situation note that the largest market for these garments is consumers in the Global North. Therefore, networks among workers and consumers provide important means to challenge these exploitative conditions. One of the groups whom Hale and Wills highlight as a 'new form of labour internationalism' is Women Working Worldwide (WWW). This UK-based NGO brings many of these garment workers together in a network to conduct public campaigns and advocacy work in Europe. Efforts such as the Clean Clothes Campaign and Ethical Trading Initiative focus on enacting change in markets and supply chains through transnational networks of garment sector workers and consumers.

Finally, fair trade and sustainable agriculture have the potential to link producers and consumers in socially responsible ways. These efforts aim to reduce labor exploitation and expand control and ownership of the means of production, especially among women. Research on fair trade and organic coffee farming finds that these initiatives can have positive effects on women farmers in Nicaragua and elsewhere in Mesoamerica (Lyon et al. 2010). Within the framework of organic and fair trade, the feminization of agriculture in this region enhances women's participation in the coffee organization. Some barriers remain, however, as women often maintain responsibility for domestic labor and have to work within traditional social norms.

Increasing emphasis on ethical farming practices in the Global South, described above, parallels the growing movement in organic and sustainable agriculture, especially among women in the US and other developed countries. In the US, Amy Trauger et al. (2010) examine the practice of 'civic agriculture,' or the social and economic strategies that benefit the strength of communities among women farm operators in rural Pennsylvania. They note that the 'rise in the number of women in farming parallels the dramatic rise in the number of organic and sustainable farming operations and farmers markets in the United States' (43). This research analyzes the motivation of women to engage in this type of farming and the articulations of gender identity in sustainable and small-scale farming. These, and other examples of ethical and sustainable economic activities, are greatly informed by feminist perspectives on globalization and community development.

This discussion examines gendered and other social aspects of women's contributions to global capitalism as consumers and workers, as well as agents of change. Women's resistance challenges the hegemonic, and often marginalizing, conditions that are evident in low-wage, exploitative jobs, human trafficking, and their subsequent removal from family and places of origin. These challenges form transgressive grassroots movements, such as those in agriculture, manufacturing, and microenterprise that are based on collaborative and sustainable practices.

Conclusion

Feminist economic geography provides an important platform to challenge conventional approaches and norms concerning social and spatial dimensions of the economy, labor, and globalization. This chapter highlights how these concepts and themes are embedded in gendered power relations and social identities, which in turn impact how people make a living. As highlighted in the vignette about Charlotte and her family in rural West Virginia at the beginning of the chapter, feminist economic geography emphasizes the intersection of the social, economic, and spatial components of where and how livelihoods take place. This perspective underscores the need to examine economic activities outside conventional frameworks of the workplace and regions to understand how work is conducted in and across the household, community, and global scales. The intersection of these scales is critical in analyses of social inequalities and uneven development. Furthermore, the global movement of capital across national boundaries overlaps with household gender relations and unequal access to resources. An excellent illustration of this is Pratt's (2012) research on Filipino domestic workers who migrate

Gendered work and economic livelihoods

to Canada and are forced to renegotiate their positions in both family and community economic and social networks.

In sum, this chapter examines these and other aspects of how work and livelihoods are gendered across different scales and over various time periods. Key geographic concepts are important in understanding feminist analyses of work, wages, divisions of labor, and neoliberal globalization. Additionally, strategies of resistance and alternative means of engaging in economic activities are central to feminist economics and geography. Household and community-based economic activities can be instrumental to individual and family survival and provide important means of empowerment for women. Likewise, globalization has provided the means to strategically organize and resist exploitation, through the mobilization of social and economic networks, local ownership of production, and ethical and sustainable agricultural models.

Recommended reading

Beneria, Lourdes, Gunseli Berik, and Maria S. Floro. 2016. *Gender, Development, and Globalization: Economics As If All People Mattered*. 2nd ed. New York: Routledge.

Gibson-Graham, J. K. 2006. *The End of Capitalism (As We Knew It): A Feminist Critique of Political Economy*. Minneapolis, MN: University of Minnesota Press.

Hanson, Susan, and Geraldine Pratt. 1995. *Gender, Work and Space*. New York: Routledge.

Kabeer, Naila, and Simeen Mahmud. 2004. 'Globalization, Gender and Poverty: Bangladeshi Women Workers in Export and Local Markets.' *Journal of International Development* 16 (1): 93–109.

Lawson, Victoria. 2007. *Making Development Geography*. New York: Routledge.

Massey, Doreen. 1995. *Spatial Divisions of Labor: Social Structures and the Geography of Production*. 2nd ed. New York: Routledge.

McDowell, Linda. 2003. *Redundant Masculinities? Employment Change and White Working Class Youth*. Oxford: Blackwell.

———. 2013. *Working Lives: Gender, Migration and Employment in Britain, 1945–2007*. Oxford: Wiley-Blackwell.

Meehan, Katie, and Kendra Strauss. 2015. *Precarious World: Contested Geographies of Social Reproduction*. Athens, GA: University of Georgia Press.

Oberhauser, Ann M. 2000. 'Feminism and Economic Geography: Gendering Work and Working Gender.' In *A Companion to Economic Geography*, edited by Eric Sheppard and Trevor J. Barnes, 60–76. Oxford: Blackwell.

Perrons, Diane. 2004. 'Understanding Social and Spatial Divisions in the New Economy: New Media Clusters and the Digital Divide.' *Economic Geography* 80 (1): 45–61.

Pratt, Geraldine. 2012. *Families Apart: Migrant Mothers and the Conflicts of Labor and Love*. Minneapolis, MN: University of Minnesota Press.

Samarasinghe, Vidyamali. 2012. *Female Sex Trafficking in Asia: The Resilience of Patriarchy in a Changing World*. New York: Routledge.

Wright, Melissa W. 2006. *Disposable Women and Other Myths of Global Capitalism*. London and New York: Routledge.

Chapter 6

FEMINIST POLITICAL GEOGRAPHY AND GEOPOLITICS

Jennifer L. Fluri

Chapter 6

Introduction

In this chapter, we examine feminist influences on political geography, geopolitics, war, and conflicts. We begin with a synopsis of gender and geopolitics, which details how the study of gender has been incorporated into political geographic analyses. This is followed by an overview of how gender roles and relations shape, and are shaped by, governments and national ideologies. Feminist geographers began by incorporating women into the study of political geography and geopolitics because they had been absent from these inquires. The inclusion of women transitioned into research on gender roles, norms, relations, and ideologies as they intersected politically with other social categories such as race, class, ethnicity, and sexuality. Much of the initial scholarship by feminist political geographers challenged existing conventional analyses, which often viewed politics as a process occurring at the macro scale of countries, governments, and international relations. By studying microscale politics, such as communities, homes, and bodies, feminist geographers effectively identify the interrelationship between the personal and the geopolitical. Examining this interrelationship involves analyzing the political aspects of daily life and intimacy, such as biological and social reproduction, and the impacts of geopolitical conflicts on intersectional gender identities and everyday life. Feminist geopolitics is considered part of **critical geopolitics**, which challenges typical and popular forms of geopolitics.

Gender and geopolitics

Feminist geopolitical research began by calling attention to the lack of women scholars in political geography *and* the absence of women (and gender relations more broadly) in the study of politics. Feminist geographers' evaluations of political geography identified that women were either missing from academic inquiry or viewed as passive, apolitical identities, and were not considered an important part of how societies allocated or distributed resources (Kofman and Peake 1990). The US second-wave liberal feminist movement's commonly used slogan, 'The personal is political,' resonated with many feminist geographers, who incorporated the study of personal issues and private spaces into political analyses. Women's organizations and movements seeking gender equality had largely been neglected by mainstream political geography, but in the 1980s and early 1990s feminist geographers established a historical account of these movements and their impacts on geopolitics internationally. Scholars

Feminist political geography and geopolitics

further examined women's participation in politics globally as voting citizens and leaders. For instance, this research identified significant gender difference in voting patterns in the US, but indistinguishable differences between the voting behaviors of men and women in the UK (Norris 1985; Pratt and Hanson 1988). In 1975, the United Nations began the International Decade of Women; at the same time, several countries continued to deny women's suffrage (e.g., Bahrain granted women suffrage in 2002; the Central African Republic granted women suffrage in 1986; Oman granted women suffrage in 2003; Palau granted women suffrage in 1979; Saudi Arabia granted women suffrage in 2015; and the UAE limited suffrage for women and men until 2006).

While achieving suffrage was a goal for many women, electoral geographic research in the 1980s made several inherent assumptions about women, work, and marital status (Kofman and Peake 1990). For example, single women were categorized based on occupation, while married women were not; this was because their positions in the workforce were assumed to be peripheral and provisional. Women's participation in leadership roles within various governments throughout the world has grown significantly in the last 30 years. However, the number of women in positions of power (relative to men) remains low. The marginalization of women from realms of political power and influence can be traced to historical political philosophies. These philosophies supported biologically deterministic representations of women: women were associated with their biology and viewed as more emotional than men. Therefore, women were commonly perceived as lacking the moral or social ability to participate in public and political life. As discussed in Chapters 3 and 5, women were most often associated with domestic spaces, child rearing, and household management, as compared to men. Although women throughout history have been actively involved in political revolutions, protests, and conflicts, their actions were minimized or underrepresented in historical accounts. Feminist historical geographers have diligently researched women's roles in politics, effectively challenging their erasure from political history. For example, Christine Sylvester (1998) traced the seemingly 'invisible' political work of women in the US during the Kennedy presidency, revealing the extensive influence of women within this administration. Feminist scholars have further examined the ways in which governments (nations, states, and nation-states) shape, and are shaped, by gender roles and relations.

Research on governments by feminist scholars has been careful to discuss the gender-specific differences between conceptualizations of 'the state' and 'the nation.' The state is a centrally organized group of institutions with the intention of controlling a place or group of people by enforcing its laws or commands, and being recognized internationally by other states. A nation is an extension of civil society (including groups and institutions) that lies 'outside the formal rubric of state parameters . . . but which both informs and is informed by them' (Anthias and Yuval-Davis 1989, 5). The nation represents the cultural and emotional aspects of a country's identity. The use of memorials, pledges, songs, and films to support a country's ideologies are examples of what makes up 'the nation.' **Nationalism** is another term used to identify practices associated with building or maintaining a nation: nationalism is the shared feelings of affinity and commitment by individuals toward a nation. The state and the nation regularly inform and create each other. Most states are also nations (i.e., nation-states), while a nation can exist without a state, such as American Indians in the US and First Nations people in Canada. These nations do not participate as states internationally or hold a seat at the United Nations. However, they identify as nations and, in many cases, have legal jurisdiction over land and people within their respective communities. Palestine is another nation without formal state recognition. Similarly, ethnically based nationalist groups, such as the Kurds or the Baloch, claim an ethno–nationalist identity while living within and across various states (Figure 6.1). While the state is more focused on the legal, political, and economic aspects of rule, the nation more often centers on various aspects of identity and history.

National interpretations of history are regularly used to shape a nostalgic connection to the nation's past; this past is used to solidify individual citizens' collective commitment to the nation in the present. Nations sometimes use gender-specific images to delineate 'proper' gender roles as part of shaping a national identity. Women's bodies are often boundary markers between existing national ideals and outside foreign influences that threaten to alter these ideals (McClintock 1993): for instance, viewing women as 'mothers of the nation' and responsible for giving birth to the ideal citizen-subject. If women mark the boundaries of the nation or homeland, when they engage with foreign concepts, ideas, modes of dress, or education, they often receive much more scrutiny than their male counterparts.

Feminist political geographers have further examined the ways in which the state is viewed as masculine (order, power, authority) and the nation as feminine (emotional, nurturing, nostalgic). For instance, women have historically been positioned as representatives of nations, while being excluded from equal or active participation in the maintenance or ruling of states (Sharp 1996,

Feminist political geography and geopolitics

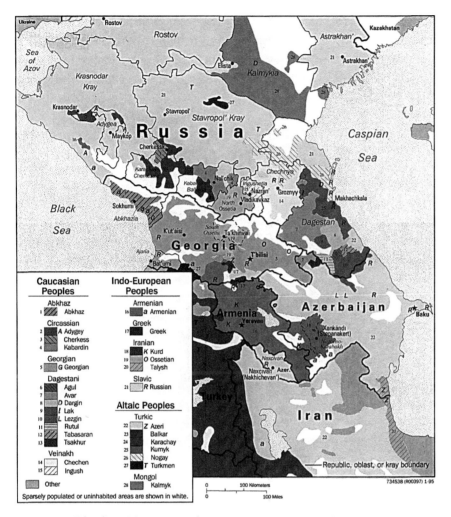

Figure 6.1 Ethnolinguistic groups in the Caucasus region. Licensed under Public Domain via Commons (https://commons.wikimedia.org/wiki/File:Ct001614.jpg)

This image is a work of a Central Intelligence Agency employee, taken or made as part of that person's official duties. As a work of the United States government, this image or media is in the public domain in the United States.

2009). Feminist political geographers both explore and challenge the ways in which access to political power has been unevenly and unequally distributed by gender, race, ethnicity, class, or sexuality. In summary, feminist political geographers seek to increase our understanding of politics from multiple scales and perspectives, including individuals and groups who have been marginalized, as well as those who are in positions of power and authority.

Despite women's marginalized position within state governance structures, women have been politically active in various ways over time and across geographic space. Women's political activism has included advocacy campaigns in various countries associated with women's reproductive rights, political interventions against gender-based violence—such as domestic abuse and rape—and increasing women's political rights. At the international scale, the United Nations Entity for Gender Equality and the Empowerment of Women (UN Women) first conceptualized the Convention on the Elimination of All Forms of Discrimination Against Women (CEDAW) at the 1975 World Conference of the International Women's Year, held in Mexico City. The UN General Assembly adopted CEDAW in 1979 (130 votes in support, zero votes against, and 10 abstentions). As this book goes to print, the US, along with Iran, Palau, Qatar, Somalia, and Tonga, are all countries that have neither signed nor ratified this convention. Many countries that have signed CEDAW have done so with reservations about specific articles in the convention (www.un.org/womenwatch/daw/cedaw/states.htm). Although contested, these attempts to stop discrimination against women at the international scale have made significant differences in political discourses and ideologies. For instance, identifying rape as a war crime, as stated by the UN Security Council Resolution 1325, outlines special care for women and children during political conflicts. At the same time, research at a more micro scale within countries, communities, and homes continues to be crucial for understanding how these mechanisms of discrimination touch down in the lived experiences of women.

Indeed, a central concern for feminist geographers has been the linkages between macro-scale policies and politics intended to improve women's rights, and the individual and community experiences of these policies. Gender-based violence has become a political issue because many feminist activists around the globe have called upon governments (and influential non-governmental agencies such as the UN) to take action against perpetrators of gender-based violence. Making private or domestic forms of violence public and political has been an integral part of several feminist movements. This form of activism challenges the idea that intimate partner or domestic violence is a private or personal issue among family members. By making this form of violence a social and political issue, feminists have sought the assistance of governing bodies to intervene on behalf of women's security. The following section provides several examples of research that addresses the spatial politics of gender-based violence in public and private spaces.

Research on feminicide in Oaxaca, Mexico, explicates the multiple factors that have contributed to violence against and murder of women in public spaces (Martin and Carvajal 2015). This research examines **structural violence** associated with gender inequality. Changing economic patterns have had an impact on gender roles, leading to an increase in the number of women occupying public spaces in the evenings, and in places considered to be inappropriate or unsafe for women. Martin and Carvajal's research underscores that gender-based violence is context-specific, continually changing, and based on simultaneous and multiple forms of political, economic, and social influence. Changing economic patterns that have resulted in more women entering the workforce have caused considerable backlash, which in several cases has been manifested in violence against female workers.

In India, the December 16, 2012 gang rape and murder of Jyoti Singh inside a moving bus galvanized protests and activism throughout the country. Many feminist groups called upon the government to take action to provide more protection for women in public space. Similar to the changing economic landscape in Mexico, women in India are increasingly working outside the home. Subsequently, more women are claiming their legitimate right to traverse public spaces at various times of the day or night without harassment or harm. Singh's rape and murder in India illustrates the urgency of movements to reinforce a woman's right to safely enter and occupy public space.

In the US, domestic and gender-based violence have become political issues. In 1994, the Violence Against Women Act became law, providing federal funding to investigate and prosecute violent crimes against women. In 2012, conservative Republicans opposed the renewal of this act; however, after a legislative dispute, the act was reinstated in 2013. Despite the political attention at the federal level, many women still routinely experience violence from intimate partners. In other situations, women have been able to extricate themselves from abusive relationships, even when courts have denied them custody because they are victims of abuse. For example, Holly Collins was the first US citizen to seek and be granted asylum by the Dutch government for herself and her children on the grounds of domestic abuse. She fled the US after a Minnesota court granted her abusive ex-husband sole custody of their two children (Nolan and Waller 2011). These examples show that despite the increase in support systems for women fleeing abusive situations, formal courts and political structures often disadvantage women who have experienced intimate partner violence.

Dana Cuomo's (2013) research on domestic violence shows how systems of assistance in the US are based on masculinist forms of protection, which are unable to attend to victims' complex security needs. 'Masculinist protection' is defined as ensuring physical protection, while marginalizing other equally important aspects of security (i.e., mental, emotional, and financial). Ignoring other forms of security by privileging masculinist forms of protection can lead to a paradoxical situation that increases (rather than decreases) victims' fear and insecurity. Cuomo links this mindset to global forms of militarism that are predicated upon similar notions of security and protection. Thus, feminist geographers are able to attend to and highlight the experiences of vulnerable bodies within the machinations of geopolitics.

The previous examples illustrate the importance of spatial and scalar political analyses of gender-based violence. Similar to federal legislation, international conventions such as CEDAW are helpful for challenging gender-based violence at a macro scale. However, many interconnected factors remain prevalent in communities and homes around the globe that require more research and understanding at the scale of communities, homes, and bodies, in order to prevent the continuation of gender-based and intimate partner violence. In addition to scale, feminist geographers, such as Lynn Staeheli, Eleonore Kofman, and Linda Peake (2004), have examined the politics of public and private spaces and individuals' embodied experiences of geopolitics.

The geopolitics of gender, race, class, and sexuality

As discussed in Chapters 2, 3, and 4, public spaces have been central to challenging federal laws and policies that disadvantage groups based on one or more aspects of their identity (such as race, gender, and/or sexuality). Feminist geographers articulate the ways in which marginalized groups utilize public and private spaces to challenge laws and policies that disadvantage them. In some cases, women activists have incorporated symbols associated with feminine biology, such as motherhood, to legitimize their protests in public spaces. This has been especially significant for resistance movements in locations where women have been perceived to have little power or legitimacy in public space.

Feminist geographers' research on the home and domestic space focuses on how these spaces can be complex political sites for both men and women. Rebel groups have used homes in order to organize political action or hide political dissidents from the state. Research shows how domestic spaces in Northern Ireland became key sites for the Irish Republican Army (IRA) to organize and strategize against British rule (Dowler 1998). By reclaiming domestic sites as places of political action, feminist geographers have helped to show the various ways in which women have participated in politics over time and across various spaces.

Feminist political geography and geopolitics

Various women's groups in the 19th and 20th centuries used motherhood, and conventional beliefs about women's maternal position in society, as a framework for political peace activism and ecofeminism. One example of the use of motherhood for political activism is Madres de la Plaza de Mayo, organized in 1977 by a group of women protesting the Argentinian government's kidnapping of their adult sons and daughters during the era known as the 'Dirty War' (1976–83). These women marched in a circle in the Plaza de Mayo in Buenos Aires, Argentina, and demanded that the government locate their missing children. Motherhood was used symbolically to situate the legitimacy of their protest, and the Plaza de Mayo is a highly public and prominent main square in central Buenos Aires. In this way, the symbolic meaning of their gendered roles as mothers and grandmothers was transferred from their expected positions in the home to a highly public and political site, in order to gain attention from the government: they brought the symbols of the private (i.e., motherhood) into the public, as a form of political action. The Madres used the respected position of motherhood to challenge the state in ways that other groups were unable to accomplish. While some of the Madres' activists were killed, their identities as mothers tended to protect them from the extensive state violence perpetuated against other groups opposing the government. In addition to resisting the state, the Madres developed international solidary networks beyond Argentina (Bosco 2001).

The use of networks by social movements has grown in recent years, with the expansion of the internet and social media technologies. The geographies of these networks exemplify what Doreen Massey (1994, 2005) identifies as 'power geometries.' **Power geometry** refers to the differentiated ways in which people experience power through flows of information, influence, and mobility. Power geometry was used to critique the idea of **time–space compression**. Time–space compression identifies the compressed amount of time it takes for people, information, or goods to travel across space due to technological advancements, such as air travel and web-based technologies. Massey uses the concept of power geometries to highlight how time–space compression is experienced differently based on one's location and socioeconomic and political situation. Cindi Katz (2004) expands on this theory by introducing the concept of **time–space expansion**, which refers to the expansion of time required to traverse space due to social, political, and economic factors. Importantly, technological advancements have not compressed time across space for some individuals because global economic and political processes have pushed these individuals to travel further from home to work, or migrate to another location for work (or to flee conflict). Thus, these individuals have experienced an expansion, rather than a compression, of time between spaces with the onset of technology and changing economic patterns.

These key interventions by feminist geographers show the importance of analyzing the interrelationships among geographic space, society, economics, and politics. When space is conceptualized as relational, we can better understand how social, economic, and political influences are complex and continually changing. These multifaceted relations are also spread across geographic spaces at various scales—including household, community, nation, state, and international.

Different experiences of time and space by individuals and groups have been further mitigated by either inclusion in, or exclusion from, spaces based on a variety of socio-political factors. Some of these socio-political factors are based on features of identity such as gender, race, ethnicity, class, and sexuality. **Identity politics** are further manifested spatially through individual and group engagements with political participation, or through being marginalized from political participation. 'Identity politics' refers to the ways in which political views are shaped by particular characteristics of individual or group identities. Feminist geographers analyze the ways in which different identities intersect to form multiple layers within various places and political situations. The concept of intersectionality has been used to understand political power, authority, influence, *and* individuals and groups on the receiving end of political policy, law, and authority.

Historically, segregation in the US demarcated spaces based on race. This racist form of politics was codified politically through law, and experienced by individual bodies in their everyday movements traversing public and private spaces. The 1896 *Plessy v. Ferguson* Supreme Court decision upheld the constitutionality of states to require and enforce racial segregation in public facilities (commonly known as Jim Crow). This legislation remained in effect until the Supreme Court ruled school segregation to be unconstitutional in the landmark 1954 *Brown v. Board of Education* decision. The 1964 Civil Rights Act banned discrimination based on race, color, religion, sex, or national origin. Individuals and groups achieved these legal changes in the US through extensive activism during the civil rights movement. Civil rights activists contentiously fought against the white establishment and racist groups, and this struggle led to the deaths of many civilians and activists, the vast majority of whom were African American men. Box 6.1 provides a detailed list of the lives and deaths of individuals involved in the civil rights movement in the US, as recorded by the Southern Poverty Law Center, a civil rights organization.

Feminist political geography and geopolitics

Box 6.1 The lives and deaths of the civil rights movement in the US

The following is a list of names of the individuals who were killed as civil rights activists or individuals caught in the crossfire:

Rev. George Lee, Lamar Smith, Emmett Till, John Earl Reese, Willie Edwards Jr., Mack Charles Parker, Herbert Lee, Roman Ducksworth Jr., Paul Guihard, William Lewis Moore, Medgar Evers, Addie Mae Collins, Denise McNair, Carole Robertson and Cynthia Wesley, Virgil Lamar Ware, Louis Allen, Johnnie Mae Chappell, Rev. Bruce Klunder, Henry Hezekiah Dee, Charles Eddie Moore, James Earl Chaney, Andrew Goodman, Michael Henry Schwerner, Lt. Col. Lemuel Penn, Jimmie Lee Jackson, Rev. James Reeb, Viola Gregg Liuzzo, Oneal Moore, Willie Brewster, Jonathan Myrick Daniels, Samuel Leamon Younge Jr., Vernon Ferdinand Dahmer, Ben Chester White, Clarence Triggs, Wharlest Jackson, Benjamin Brown, Samuel Ephesians Hammond Jr., Delano Herman Middleton, Henry Ezekial Smith, and Dr. Martin Luther King Jr.

Activity

Go to the Southern Poverty Law Center's Civil Rights Martyrs webpage at www.splcenter.org/what-we-do/civil-rights-memorial/civil-rights-martyrs to read further about the lives and deaths of these individuals. Consider the ways in which the civil rights movement shaped racial politics in the US and identify the contemporary problems with racism in the US today.

White supremacists and white nationalist groups, such as the Ku Klux Klan, perpetuated (and continue to perpetuate) political and vigilante violence (murder, torture, beatings, etc.) in an effort to intimidate and challenge those working toward racial equality. Women have been integrated into white supremacist organizations based on their association with biological and social reproduction to ensure so-called 'racial purity' (Fluri and Dowler 2003). The spatial segregation during the Jim Crow era focused mainly on public spaces, while African American women experienced racism in the private spaces of white homes through their work as housekeepers and caregivers. Today, white supremacist groups are no longer mainstream political organizations, but the rise of populist right-wing groups is occurring in the US and other countries, and the racial marginalization of non-whites persists through various forms of institutionalized racism and structural violence. **Institutionalized racism** continues today through mass forms of incarceration that disproportionately target African American men (Alexander 2012).

Apartheid in South Africa is another example of spatial segregation based on race and codified politically through law. South African racial apartheid was an outgrowth of British colonialism and codified in law by the white minority rulers of the National Party in 1948, remaining in effect until 1994. Civil rights activists in South Africa fought against apartheid for several decades, during which time many activists were killed or imprisoned. Former South African President Nelson Mandela was an anti-apartheid activist who served a 27-year prison term. After the end of apartheid, Mandela was elected President of South Africa (1994–9) and his government worked to dismantle the legal legacy of apartheid and institutionalized racism in South Africa.

Research by Katherine McKittrick (2006, 2011) and McKittrick and Clyde Woods (2007) examines the politics of gender and racism by elucidating Black geographies through in-depth analyses of Black diasporic histories. McKittrick's research investigates the multiple scales and spaces of the transatlantic slave trade, from the transport ships to the auction blocks and plantations. She identifies the ways in which Africans were socially, politically, and economically marked as commodities, their bodies repetitively violated in the pursuit of profit. In this context, McKittrick (2011) asks us to consider a 'black sense of place,' which she defines as a process of understanding 'historical and contemporary struggles against practices of domination and the difficult entanglements of racial encounter' (949). She cautions that resistance to racism in the Americas is not the only defining marker of a 'black sense of place' as it is linked to multiple and vacillating geographic and historic contexts.

The preponderance of **relational violence** has constructed the condition of being Black through continuous sites of struggle. Ruth Wilson Gilmore's (2007) research challenges conventional racial politics that bifurcates understanding by placing people into us/them categories. She challenges her readers to conceptualize human relations through a corporeal worldview by radically challenging contemporary forms of common sense about racial and sexual classifications that profit from categories of separation and difference. Her research challenges the existing framework of race, gender, and sexuality by destabilizing insider/outsider categories and seeking a more thoughtful and fruitful examination of the sites through which cooperative human efforts take place. Additionally, scholars focus on the discipline of geography's white supremacist history associated with imperial conquest, while others argue that geographers need to work harder to ensure that the discipline becomes more open, inclusive, and diverse. Minelle Mahtani (2014) uses the term 'toxic geographies' to describe the marginalization experienced by scholars of color within the overwhelmingly white discipline of geography: 'In geography toxicity has sustaining, long-term implications not only for the lives of scholars of color, but it also impacts the scholarship on race and difference' (360). Several critical geographers have called for more diversity within the discipline and scholarship that critically examines the spatial experiences of race and gender. Increasing the diversity of scholars and scholarship will thereby influence the ways in which knowledge is shaped and shared. The politics of race and gender internationally, and in the US specifically, illustrate various sites of struggle. As feminist geographers have shown, political rights require multiple forms of spatial inclusion and must be articulated through activism, legal and policy changes, and in the ways in which we produce knowledge through academic disciplines. In addition to race and gender, the political marginalization or inclusion of certain groups has further been demarcated by sex and sexuality, including efforts to control biological reproduction.

Sex and sexuality

Just as feminist geographers have analyzed politics at the scales of communities, homes, and bodies, explicating the ways in which private spaces can be highly political, the 'everyday' (in addition to the exceptional and dramatic) has become an integral part of feminist geopolitics. Feminist political geographers show how vulnerable bodies attempt to 'negotiate and transform the geopolitics they both animate and inhabit' (Dixon and Marston 2011, 445). The concept of **intimate geopolitics** seeks to disrupt the customary divisions between global/local, familial/state, and personal/political by showing how they are mutually constituted, affected, and integrated (Smith 2012; Pain and Staeheli 2014). 'Intimacy' is defined as a set of spatial relations with an emphasis on homes or private spaces, a method of interaction that can stretch from interpersonal to faraway, and a set of practices connected to the body. Intimate geopolitics represents the ways in which political control, violence, and security are enacted on and through bodies.

Changing political ideologies and policies can have an impact on the intimate relationships of people living in the same geographic area who are associated with different political categories. Sara Smith's (2009, 2011, 2012,

2013) research explicates the ways in which Hindu nationalism in India had an impact on the intimate practices of marriage and biological reproduction. For instance, Buddhist and Muslim inter-marriage had previously been accepted in Leh, Ladakh, India. However, after the rise to power of the Hindu right-wing Bharatiya Janata Party (BJP) in India, inter-religious marriages became taboo and were no longer accepted within the community. For example, this illustrates the domestication of geopolitics, or the 'bringing of geopolitics into the home and into the personal spaces of the body, marriage, and intimacy' (Smith 2009, 212). The Love under Apartheid project in Palestine is an example of how love becomes a counter-geopolitics through storytelling (Marshall 2014). These stories seek to confront and attempt to dismantle the spatial divisions of Palestine in Gaza, the West Bank, and Jerusalem by providing intimate knowledge about Palestinians while appealing to universal human rights discourses.

Critical geographies of identity politics call into question the privilege afforded to certain bodies and the marginalization of others. Political 'othering' has often been used by states to mobilize certain citizens in support of the nation. In these cases, policies that seek to control reproduction have been used in an effort to reproduce the 'wanted' citizen and limit the biological reproduction of the 'other.' Tamar Mayer's (2012) research illustrates the multiple ways in which the Israeli state has encouraged Jewish women's biological reproduction. In what Mayer refers to as the 'womb wars,' women's bodies have become a reproductive battleground through state-based polices that attempt to increase Jewish, and limit Palestinian, reproduction. Women are often perceived, then, as physical or symbolic reproducers of the nation. This intersects with racial and ethnic divisions within countries in an effort (as seen in the above examples) to reproduce a particular citizenry based on various intersectional identities defined by the nation-state.

Feminist geographers focus on the sites of these interactions and their divergences. For example, Rachel Pain and Susan Smith (2008) use the visual metaphor of the double helix to examine complex political dynamics: one strand representing everyday lives and another international geopolitics, each intersecting with the other through numerous entanglements. The body is another way in which the intimacies of geopolitics are examined. Bodies carry particular representations of citizenship, privilege, and the lack thereof. Privileged bodies have been used as a method of protecting other bodies that have been targets of violence. Transnational solidarity groups in Guatemala and Colombia, for instance, have privileged volunteers accompany peace activists to protect them from harm while moving in public spaces. This has been labeled a form of 'proxy citizenship' (Henderson 2009). Accompaniment is an example of proxy citizenship because the individual operating as an accompanier to provide 'protection' is generally a non-citizen of that country, while carrying citizenship status from a more powerful country. Therefore, his/her body acts as a form of protection for the individual who is under threat. The accompaniers' bodies and foreign citizenship serve a proxy function for the targeted person. Research on accompaniment in Colombia identifies this (and related forms of peace activism) as 'alter-geopolitics' and suggests that more studies should be done on peace work, since a significant amount of political geography focuses on political conflict and war (Koopman 2011). Feminist geographers have further examined how governments incorporate bodies and intimate relationships as part of border management practices.

Lauren Martin's (2011) research explicates how the US government's construction of family and childhood detention facilities has been used to criminalize migrant laborers traveling with children. In her analysis, she shows how the intimate relationships between parents and children are disrupted, through detention practices and the representation of laborers as criminals rather than caregivers. In this way, the complexities associated with families and economic needs at an intimate scale are often ignored through a strict policing of the US–Mexico border. Feminist geographers' insights are important in this context because of the ways in which they examine how national political action, discourse, and policy affect our bodies and our most intimate activities.

Political determinations of both sexual relationships and marriage practices (i.e., which bodies are permitted to have sex and/or marry, and with whom) exemplify how governments attempt to control our most intimate activities. For example, same-sex marriage was illegal in most US states prior to the June 26, 2015 decision by the Supreme Court, which identified any ban on same-sex marriage as unconstitutional. While this was a landmark case that dramatically shifted government regulation of marriage, other restrictions on marriage remain. For example, polygamy and polyandry remain illegal in the US, along with alternative conceptualizations of families that do not fit the two-partner model. Social and cultural barriers still remain in areas where conservative Judeo-Christian notions of sexuality dominate.

The International Lesbian, Gay, Bisexual, Trans and Intersex Association (ILGA) identifies 75 countries that have criminal laws against sexual activity by lesbian, gay, transgender, intersex, or bisexual people. In some countries where polygamy is legal, gay and lesbian marriage (along with polyandry) remains illegal. The political geographies of sex and sexuality call into question the role of governments in determining citizens' sex partners, which citizens are seen as desirable, and which are marginalized based on one or multiple aspects of their identities. Sexual citizenship has become an area of focus for feminist geographers, who examine the ways in which **heteronormative sexuality** is politically identified and challenged.

Jasbir Puar's (2007, 2013) research critiques the co-optation of sexual rights discourses to garner support for the use of military force. Her work examines US military personnel's torture of inmates in Abu Ghraib prison in Iraq through an analysis of the series of photographs that were publically leaked. In these photographs, US service personnel are seen torturing Iraqi prisoners with electric shocks, water boarding, forcing prisoners to wear dog leashes and 'walk' on hands and knees, sexually abusing female prisoners, and forcing male prisoners to perform or simulate homosexual sex acts. Puar states that 'both the justifications and the admonishments provoked by the Abu Ghraib photos rely on **Orientalist** [prejudicial] constructions of Muslim male sexuality as simultaneously excessively queer and dangerously premodern. The discursive field produced around Abu Ghraib enlists homonormative U.S. subjects in the defense of "democratic" occupation' (Puar 2013, 336).

Puar's research on these instances of torture by US military personnel highlights the ways in which Muslim men have been placed into narrow categorizations that reproduce racist, sexist, and sexualized stereotypes of Muslim men. The Iraqi prisoners were identified as sexually repressive and pre-modern in order to frame Islam as sexuality-intolerant and to divert public attention away from the US' illegal invasion of Iraq. This invasion was based on false information that Iraq was involved in the September 11, 2001 terrorist attacks on the US, and that it possessed weapons of mass destruction—both of which have been identified as untrue assumptions that were gained from torturing Guantanamo Bay prison inmates. Guantanamo Bay and Abu Ghraib prisons are militarized spaces, and feminist geographers examine gendered dimensions of lived experiences within such militarized spaces.

Militarized spaces and bodies

Territory is gendered, which means that (1) spaces can be demarcated by gender, (2) territory can be symbolically associated with one gender or another (i.e., motherland, fatherland), and (3) men's and women's bodies have been differentially associated with territorial conquests along with land and resource struggles. Bodies are implicated as part of militarization. For example, men are often more associated with the protection of the nation through various constructions of masculinity. Combat masculinity or militarized masculinity represents a specific form or construction of masculinity, associated with war-related violence.

Matthew Farish (2010) examines the ways in which male bodies have been subsumed into different forms of political violence as soldiers. Farish's analysis of Cold War military bodies includes investigating the ways in which the US military 'tested' its own soldiers. The following quote from his book exemplifies the ways in which soldiers' bodies were incorporated into the machinations of war:

> The combination of eyes, ears, brains and muscles formed the 'indispensable tools of war,' and military equipment was only an extension of bodily ability. By establishing this hierarchy and prioritizing bodies . . . they were also solidifying the concept of 'man-machine units,' later expanded in what was called human factors research such that the truly effective solider was a combination of technological design and bodily training.
> (153)

Soldiers represent national strength and courage, while their bodies and lives are simultaneously expendable in the service of national protection and defense.

Cynthia Enloe's (1988, 1993, 2000) extensive research on militarism has had significant influence on feminist political geography. She examines the way in which a 'loot, pillage, and rape' approach has been integral to war and militarism. As stated in Nicole Detraz's (2012) study of gender and security:

> The strategy of systematic rape has taken different forms depending on the conflict, but is evident as far back as the Crusades, to the First World War with the rapes committed as the Russian Army marched to Berlin, to the Second World War with the sexual slavery of Jewish women for Nazi soldiers and the enforced institutionalized rape of 'comfort women' by the Japanese army, in the Bosnian war with the Serb rape camps, and in the Rwandan genocide with the rape of Tutsi women, sometimes instigated by Hutu females . . . It is estimated around 200,000 women . . . were raped during the nine-month conflict in which East Pakistan broke from West Pakistan to become Bangladesh.
> (39–41)

The use of rape as a weapon of war was first identified as a war crime by the UN in 1993, and the International Criminal Tribunal for the former Yugoslavia (1993) and the International Criminal Tribunal for Rwanda (1994) recognized rape as a crime against humanity. UN Security Council Resolution 1325 requests that member states provide special provisions and protections for women and girls during armed conflict; however, critiques of this resolution question the ability of the UN to orchestrate or enforce these provisions (Shepherd 2008). Gender has also been used as a military tool in the most recent US-led conflicts in Afghanistan and Iraq. Female Engagement Teams are one example of how gender became incorporated in US military actions in Iraq and Afghanistan, which is elaborated upon in Box 6.2.

Box 6.2 Female Engagement Teams (FETs)

Female Engagement Teams (FETs) began as all-female marine units (and were later included as part of army units) who would be taken into conflict areas under the protection of all-male combat units in Iraq and Afghanistan. The female soldiers were used to help draw out men in villages who were expected to be curious about them. Female soldiers were perceived as less threatening to local communities, based on patriarchal gender hierarchies. Also, FETs would offer medical services to communities as a method of stimulating interaction and engagement between the soldiers and the village populations. As women, the FETs were often allowed by local Afghans and Iraqis to enter homes. Male soldiers were not welcomed into local homes because they were, by definition, perceived as a threat based on their gender (and a long history of rape as a war weapon). Gendered demarcations of space in these villages associated men with public space and women with private/domestic spaces. Thus, US forces, by way of the FET program, manipulated these gendered divisions of space for geopolitical military purposes.

The FETs were part of the US counterinsurgency (COIN) program, which attempted to 'win hearts and minds' of villagers in order to gain their support and challenge the influence of local insurgent groups. FET access to domestic spaces was also intended to gather information from the perspectives of local women. Several FET soldiers also challenged the US military ban on women in combat, which was recently lifted, with specific restrictions on the number of women allowed in combat.

Feminist political geography and geopolitics

Enloe further examines how militarism has led to a 'naturalization' of violence as an expected (rather than exceptional) response to inter-state tensions and ineffective diplomacy. Militarism is associated with particularly strict codes of masculine behavior and connected to modernity and state sovereignty. In this way, she has shown how militarization has come to define peoples' lives. Box 6.3 provides an example of everyday militarism in Afghanistan.

Box 6.3 Everyday militarization

Figure 6.2 An ice cream seller at Qarghar Lake, Afghanistan. Reproduced courtesy of Jennifer Fluri.

On a research trip to Kabul, Afghanistan, in 2015, Jennifer Fluri and her collaborator Rachel Lehr were speaking with their Afghan friend and colleague, Najibullah. They were joking about the sounds of the ice cream sellers that filled the noisy city with the unrelenting and repeated 'happy birthday' tune (Figure 6.2). Men push freezer-cooler carts on wheels, the tune signaling their proximity as they attempt to entice individuals to purchase ice cream. Although these carts are ubiquitous throughout the city (especially during the warm June days of early summer), they rarely observed anyone purchasing ice cream from these sellers. Najibullah stated that he would not allow his children to go near the ice cream sellers. When they asked why, he explained that, to him, the freezer boxes looked large enough to conceal a bomb. He viewed the ice cream sellers as a site of potential violence.

In a city besieged by multiple forms of violence from US-led international forces, local insurgents, and the Taliban (and at that time regular acts of bombing and other violence as part of the Taliban's 2015 spring offensive), everyday acts—such as selling ice cream—had become suspect. Something as ordinary as buying an ice cream from a pushcart had become marked as a potential site of violent disruption. Najibullah had grown up in Kabul and had lived in the city during each phase of nearly 40 years of war and related violence. Violence was part of, and embedded into, his everyday life, rendering the ice cream seller's cart a militarized object. In many respects, this example reflects what Cynthia Enloe calls 'militarized thinking' (Enloe 2000), particularly by individuals (like Najibullah) who have never been soldiers or operated a lethal weapon. This story is also an allegory for highlighting the invisible experiences of non-combatant and non-military men living and working in conflict zones. Simply because the majority of soldiers and combatants in most conflicts are male does not mean that all men are fighters, and many civilian (non-combatant/non-soldier) men are themselves victims of violence (Carpenter 2006). Therefore, while women and children's vulnerabilities are heightened and increasingly highlighted during conflicts, non-combatant men are rendered invisible.

Activity

The Twitter hashtag #AfghanistanYouNeverSee is one of the ways in which individuals living or traveling within Afghanistan have attempted to challenge journalistic reporting that focuses predominantly on conflict and destruction.

1. Search this hashtag and view the various images and discussions on Twitter.
2. How do these images and stories differ from your knowledge or understanding about Afghanistan?

The militarization of everyday lives is one of many examples of the geopolitics of daily life. As discussed earlier in this chapter, feminist political geographers have challenged the assumption that politics occur only at the scale of the nation, state, or through international relations, by examining the

Feminist political geography and geopolitics

ways in which politics affect, and are affected by, everyday life. Rachel Pain (2014) describes the ways in which fear is manipulated at different scales to incite violence, control populations, and devise security regimes. Her research highlights the inter-linkages among domestic violence, political conflict, war, and terrorism. She states that 'intimate violence has congruence with international conflict, its emotional dynamics and tactics strongly resonate with the conduct and psychology of warfare' (352). For example, during times of heightened political conflict or war, domestic violence increases among military personnel.

Part of the US-led invasion of Afghanistan in 2001 involved the discourse of 'saving Afghan women.' While this discourse was effective in garnering public support in the US for continued military operations in Afghanistan, many Afghan feminist activists and international feminist scholars viewed the actual 'saving' of Afghan women as a co-optation of women's rights for alternative political purposes. Research on gender and geopolitics in Afghanistan reveals that 'saving women' operated effectively as a powerful discourse but was not extended to all women living and working in this conflict zone (Fluri 2011). International female (development, aid, and service) workers, for example, experienced various forms of sexual misconduct, harassment, and abuse from international men (development workers, contractors, and logistics and private security personnel). International sex work and the proliferation of brothels in Afghanistan occurred alongside the massive influx of international workers as part of the US-led military, aid, and development interventions beginning in 2001 (Fluri 2009, 2011). Enloe (1993, 2000) has examined the proxy bodies of sex workers who are expected to provide sexual services to male soldiers in conflict zones. Enloe (2000) explicates the linkages between militarization and sex by identifying four key aspects of wartime economies associated with the proliferation of sex workers in areas beset upon by large numbers of (mostly male) military personnel:

1) When large numbers of local women are treated by the government and private entrepreneurs as second-class citizens, a source of cheapened labor, even while other women are joining the newly expanded middle class. 2) When the foreign government basing its troops on local soil sees prostitution as a 'necessary evil' to keep up their male soldiers' fight morale. 3) When tourism is imagined by local and foreign economic planners to be a fast road to development. 4) When the local government hosting those foreign troops is under the influence of its own military men, local

military men who define human rights violations as necessary for 'national security.'

(70–1)

Soldiers and civilians experience the militarization of everyday life through the ways in which their bodies are incorporated into the systems of military machinations and spaces. From the use and abuse of soldiers, to rape as a weapon of war, and gendered bodies as a reason to enact or sustain conflict, bodies are central to military maneuvers, machinations, and manipulations.

Conclusion

In this chapter, we have examined the impact of feminism on political geography and geopolitics. This discussion began with the inclusion of women as scholars and as part of political processes and analyses, and moved on to the ways in which gender shapes, and is shaped, by politics. Gender intersectionality was explored in this chapter by examining the ways in which governments have controlled (and continue to control) sex, sexuality, and biological reproduction. These and other forms of intimate geopolitics highlight the interrelationship between geopolitics and the everyday lives of communities and individuals. Thus, scale has been a central concern for feminist political geographers, who argue for analyses at multiple scales of inquiry from the macro (international, regional, and national) to the micro (communities, homes, and bodies). By researching these everyday geographies, feminist geographers elucidate the ways in which our daily lives influence and are influenced by politics.

Recommended reading

Dixon, Deborah P., and Sallie A. Marston. 2011. 'Introduction: Feminist Engagements with Geopolitics.' *Gender, Place & Culture* 18 (4): 445–53. doi:10.1080/0966369X.2011.583401.

Dowler, Lorraine, and Joanne Sharp. 2001. 'A Feminist Geopolitics?' *Space and Polity* 5 (3): 165–76.

Fluri, Jennifer L. 2011. 'Armored Peacocks and Proxy Bodies: Gender Geopolitics in Aid/Development Spaces of Afghanistan.' *Gender, Place & Culture* 18 (4): 519–36.

Hyndman, Jennifer. 2004. 'Mind the Gap: Bridging Feminist and Political Geography through Geopolitics.' *Political Geography* 23 (3): 307–22.

Koopman, Sara. 2011. 'Alter-Geopolitics: Other Securities Are Happening.' *Geoforum* 42 (3): 274–84. doi:10.1016/j.geoforum.2011.01.007.

Martin, Patricia M., and Nohora Carvajal. 2015. 'Feminicide as "Act" and "Process": A Geography of Gendered Violence in Oaxaca.' *Gender, Place & Culture,* 23 (7): 1–14. doi:10.1080/0966369X.2015.1073693.

McKittrick, Katherine. 2006. *Demonic Grounds: Black Women and the Cartographies of Struggle*. Minneapolis, MN: University of Minnesota Press.

Sharp, Joanne. 2009. 'Geography and Gender: What Belongs to Feminist Geography? Emotion, Power and Change.' *Progress in Human Geography* 33 (1): 74–80.

Smith, Sara. 2012. 'Intimate Geopolitics: Religion, Marriage, and Reproductive Bodies in Leh, Ladakh.' *Annals of the Association of American Geographers* 102 (6): 1511–28. doi:10.1080/00045608.2012.660391.

Chapter 7

ENVIRONMENTAL STRUGGLES ARE FEMINIST STRUGGLES
FEMINIST POLITICAL ECOLOGY AS DEVELOPMENT CRITIQUE

Sharlene Mollett

Introduction

> We must undertake the struggle in all parts of the world, wherever we may be, because we have no other spare or replacement planet. We have only this one, and we have to take action.
>
> (Berta Cáceres, *Guardian*, May 2015)

In March 2016, Berta Cáceres, co-founder of the Lenca organization COPINH (Council of Indigenous Peoples of Honduras) was assassinated in her home. Prior to her death, Berta had received the Goldman Environmental Prize—the world's leading environmental award—for her leadership and opposition against the Agua Zarca Dam (Figure 7.1). This dam and accompanying reservoir were to be located in western Honduras, along the Gualcarque River Basin and the area of Rio Blanco, situated between the departments of Intibucá and Santa Barbara, a region that overlaps with the ancestral territories of the Lenca peoples. Berta and COPINH opposed this controversial hydroelectric power project primarily because dam construction threatened to desiccate Lenca water sources. For COPINH and the communities it represents, this project's construction and operation, run by a Honduran corporation DESA (Desarrollos Energéticos S.A.), made land loss and environmental degradation imminent, until most recently when key investors for the Agua Zarca Dam project began to permanently withdraw financial support.

Figure 7.1 Berta Cáceres, a Lenca and Women Human Rights Defender from Honduras. She was assassinated in her home in March 2016 for leading dissent against the Agua Zarca Dam. Photograph by David Shankbone. Reproduced courtesy of the Goldman Environmental Foundation.

Environmental/feminist struggles

Chapter 7

Today, while the future of the dam is uncertain, due to indigenous, environmental, and human rights activism, Berta's murder exemplifies a growing violent trend in Mesoamerica (Mexico and Central America). The region's violence, while overwhelmingly targeting indigenous peoples assumed to be in the way of unsustainable and environmentally destructive state and elite development projects, is also strikingly aimed against women activists leading environmental and social struggles. In Central America, these women argue that environmental destruction and social displacement impinge on the human rights of their communities; collectively, they are known as 'Women Human Rights Defenders' (WHRDs) (IM-Defensoras 2015). WHRDs like Berta, along with men, lead their communities against state and elite repression and extractive practices associated with mining, logging, and oil production, as well as infrastructural projects like dams and hydroelectricity that pose exponential harm to water, land, and air quality. But the economic interests of the state and elite are not the only motivations behind a common barrage of death threats and persecution against WHRDs; racial discrimination, patriarchal norms, and male dominance within their societies also shape such conflicts.

Berta Cáceres and her struggle for environmental justice and human rights for the people of Honduras is a story with multiple and overlapping struggles illuminating how environmental conflicts, feminist contests, and the associated forms of violence, sexual slavery, torture, rape, and feminicide are entangled. Such entanglements of power around race, patriarchy, and capitalism are growing foci within the subfield of feminist political ecology.

Hence, this chapter is designed to introduce students to **feminist political ecology (FPE)** and offers a partial view of the subfield by including scholars who may or may not use the FPE label, but whose research shares feminist political ecological sensibilities. As a cultural geographer and feminist political ecologist, I highlight scholarship that is attentive to the multiple kinds of power that are shaping gender subjectivity. This chapter is not meant to be exhaustive. Instead, throughout the chapter I offer detailed examples from a few case studies so that students and instructors can think with, discuss, and enliven classroom debate through details and dialogue in hopes of sparking student research interests that move beyond these examples. Such case studies and their embedded power complexities offer familiarity for students looking to conduct similar kinds of research, whether online, in their local communities, in their university libraries, and/or in the field.

Feminist political ecology

Feminist political ecology (FPE) emerged as a subfield of political ecology in the early 1990s. **Political ecology** is an eclectic field of inquiry within geography (and other disciplines in the social sciences) and represents a framework that investigates the entanglement of political economy with varied meanings of ecology, as a way to understand environmental change. This field covers a large array of topics such as poverty, environmental degradation, land conflicts, biodiversity conservation, extraction, food (in)security, more-than-human natures, the enclosure of the commons, and the role that capitalism plays in the shaping of nature(s) (among many others). At its center lies a commitment to social justice and making visible the way in which societal inequalities are 'structural.' Power structures are woven into the production of space and the distribution of nature. Political ecology is a useful framework for illuminating how power relations shape access to resources, the meanings of nature, and the distribution of rights to nature and natural resources. This plural field of geographic inquiry, shaped by a variety of ideologies, is a conceptual framework that examines 'the complex relations between nature and society through a careful analysis of . . . access and control over resources and their implications for environmental health and sustainable livelihoods' (Watts 2000, 257).

During the 1990s, FPE emerged as a response to a disproportionate focus on male environmental stakeholders and what feminist geographers insisted was an inadequate attention to the more feminist objectives, strategies, and practices in the field of political ecology (Rocheleau et al. 1996; Jarosz 1991; Carney 1996). In 1996, Dianne Rocheleau, Barbara Thomas-Slayter, and Esther Wangari introduced their influential volume *Feminist Political Ecology: Global Issues and Local Experiences*. In this work, Rocheleau et al. are explicit in arguing that FPE must employ gender as 'a critical variable in shaping resource access and control, interacting with class, caste, race, culture and ethnicity to shape processes of ecological change' (4). In their volume, the authors highlight the lived experiences of grassroots environmental movements that have global visibility, for instance the Chipko Movement (Figure 7.2), as well as more hidden and diffuse gendered resource struggles that take place within the household and across communities. Rocheleau et al.'s main concern focuses on survival and the rights to live and make livelihoods in robust environments; the duty to protect the environment from extraction, pollution, contamination, and degradation; and finding ways to 'restore' and 'rehabilitate' communities and their environments after degradation (often through various forms of activism).

Environmental/feminist struggles

Intersectionality and feminist political ecologies

Like the broad subfield of feminist geography, FPE approaches are further enriched by engagement with intersectionality. This serves to recognize the complex power relations shaping the lives of people in the context of international development in the Global South. As such, feminist geographers have recently argued for a **postcolonial intersectionality** (Mollett and Faria 2013; Radcliffe 2015). As Mollett and Faria (2013) insist, 'postcolonial intersectionality attends to how patriarchy and racialized practice are consistently rooted in a postcolonial genealogy that grounds gender and racial exclusions, privileges and struggles within international development processes' (120). This concept is useful for demonstrating how humans and non-humans are consistently marked by difference. But most specifically, to date, post-colonial intersectionality employs Mohanty's critique (see below) against the construction of a 'Third World woman' and highlights 'a grounded and spatially informed understanding of patriarchy constituted in and through racial power' (Mollett and Faria 2013, 120).

These concerns and insights, however, come together and build upon prior research in the area of gender, environment, and development.

Figure 7.2 The Chipko Movement. Members of this movement hugged trees as a symbolic and material expression of the need to protect the forest. Reproduced courtesy of Panos Pictures. (www.panos.co.uk/preview/00034807.html?p=1)

FPE influences

Ecofeminism and feminist environmentalism

Ecofeminism is the acknowledgement that there are significant connections between the treatment of women, people of color, and the poor, on the one hand, and the treatment of non-human nature on the other. Ecofeminists argue that these connections are salient. Like feminism, there is not one ecofeminism, but a variety of ecofeminisms shaped by many influences, from liberal feminisms, to Third World feminisms. Indeed, non-human nature and the domination of nature are feminist concerns. This philosophy and practice is commonly understood through the celebration of Earth Mother figures and an essentialization of the role of women in knowing and caring, made popular around the globe largely through the scholarship and activism of

Figure 7.3 Women farmers in Rajkot, Gujarat, India. Agricultural workers returning to their homes by motorbike and trailer at dusk. Despite the recent focus on industrialization, agriculture remains a vital part of the Indian economy and employs roughly half the country's workforce. Photograph by Tim Smith. Reproduced courtesy of Panos Pictures. (www.panos.co.uk/preview/00218994.html?p=1)

Environmental/feminist struggles

Vandana Shiva. Vandana Shiva is the iconic symbol of ecofeminism around the world. She is the founder of Navdanya (Nine Seeds) in India. Navdanya is a women-focused social movement that aims to protect biological and cultural diversity. While widely respected, however, Shiva's work was challenged as women's so-called *innate* ties to nature were discredited by many feminists who contested such essentializations about 'women' as a homogenous group. Nonetheless, today she remains an important global social justice activist and proponent of organic farming, seed saving, and food security, and advocates for women's participation in agriculture as essential for planetary well-being.

Feminist environmentalism, based both in the US and India, offers another way to understand the politics of gender and the environment. In the US, feminist environmentalism grew from wider debates made visible through feminist organizing, and a range of movements from anti-militarist and anti-pollution movements. Feminist environmentalism came to be understood as 'a fusion of powerful analytical and paradigmatic challenges and activist energies' (Seager 2003, 947). In India, feminist environmentalism is best exemplified by the work of Bina Agarwal and her instructive text, *A Field of One's Own* (1994). In this work, Agarwal focuses on the multiple relations of power that subjugate women in their private space (households) and public space (villages). She argues that because of ongoing environmental loss, women are compelled to spend more time doing daily chores, leaving less time for agriculture and food production, which reduces nutritional intake and leads to increased illnesses among women and their families. With a focus on the political and economic effects of environmental change in rural India, her approach situates gendered aspects of environmental change in the interwoven relations of property, patriarchy, and labor, and serves as an important counter-narrative to ecofeminist essentialisms.

US and Indian feminist environmentalisms align through protests against 'a society and world economy organized for the profit of a small number of white men [which] has created the conditions for widespread unemployment, violence at home and in the streets, oppression of third world peoples, racist attacks, inadequate food, housing and health care, and finally the ecological devastation of the earth' (Women and Life on Earth 1979, cited in Seager 2003, 948).

Feminist political economy

Research in the field of feminist political economy also overlaps and shapes thinking within feminist political ecology. Since the 1970s, feminist political economy scholars have focused on making visible the gendered hierarchies embedded in processes and practices of international development. Ester Boserup's pathbreaking study *Women's Role in Economic Development* (1970) interrogated the way in which development theories, practices, and policies elided and subjugated women's roles as farmers and wage earners. In reference to Sub-Saharan African countries, when typically women were central to family agricultural systems, land inheritance, and cultivation decisions, colonial and national officials assumed women's contributions were made *only* inside households. These assumptions meant that agricultural planning as part of development programs imagined and promoted men's labor as the most efficient and productive (Boserup, cited in Wilson 2015, 805), while women's contributions were deemed outside production.

Thus, **Women in Development (WID)** approaches launched a campaign that argued for the inclusion of women in development thinking and practice. Advocates cogently argued that women could be as 'efficient' and 'productive' as men (Wilson 2015). Relatedly, their exclusion was the result of sexist stereotyping by development planners and officials in state development agencies. In response, a discourse of efficiency unfolded and was adopted broadly by international development organizations throughout the 1980s. A focus on WID was based on assumptions that development would be more 'efficient' if women's resources were fully 'utilized.' Rather than directly countering gender discrimination as a structural form of power outside development, women's participation in the market was seen to be synonymous with gender equity and exalted for leading to higher incomes, employment, and greater productivity among women (Kabeer 1994; Wilson 2015).

However, rather than celebrate women's increased access to the market, **Women and Development (WAD)** critics, largely influenced by Marxist and socialist feminisms, theorized that capitalist production, at both the nation-state and international scales, was made possible through the ongoing subjugation of the Global South, and of women. In fact, according to Marxist feminist political economists, women and the Global South 'are treated as if they were means of production or "natural resources" such as water, air, and land . . . the relationship between them is one of appropriation' (Mies et al. 1988, cited in Rai 2013, 34). Maria Mies (1982, 1986) 'explored how both local gender relations which constrained women's mobility and global capitalist characterizations of women's work as unproductive "housework" generated expanded profits for capital on a world scale' (Wilson 2015, 806). While such critiques culminated under the leadership of **Development Alternative with Women for a New Era (DAWN)**, WAD pushed for more detailed and nuanced attention to geographies and histories of place to counter many of the celebratory narratives of capitalism that both ignored and relied upon patriarchal power relations and women's subjugation (Kabeer 1994; Kandiyoti 1988; Sen and Grown 1985).

As a critique to both WID and WAD, **Gender and Development (GAD)** efforts look to rethink 'women' as a homogenous category by highlighting the influences of difference such as age, race, sexuality, marital status, religions, ethnicity, and regions. GAD explores how development practice reshapes gender relations. This framework is underpinned by the notion that women are important agents of change. Moving away from WID and WAD programs, GAD efforts seek not to see 'women' as a uniform category, but as groups of women with overlapping priorities. Thus, the thinking behind GAD approaches is that 'women-only' policies fail to acknowledge gender as a socially constructed category. Thus, GAD efforts aim to 'not only integrate women into development, but [also to] look for the potential in development initiatives to transform unequal social-gender relations and to empower women' (Canadian Council for International Cooperation 1991, cited in Bhavnani et al. 2003, 823).

GAD approaches also share some priorities with the **Women, Culture, and Development (WCD)** approach. This approach argues

> that production and reproduction cannot be separated in the lives of most women . . . that is women all around the world are usually expected to be in jobs, where the necessary skills (and limitations) for 'women's work' are derived from ideological notions of women's abilities as well as from the work of women that contributes to household needs, . . . of children, cooking food, and so on. What many have referred to as the 'double shift.'
> (Bhavnani et al. 2003, 7–8)

WCD approaches very much align with the work of Bina Agarwal (above) and highlight the way in which land rights were, and remain, a significant development priority for rural women in the Global South. Rethinking women's rights to land clarifies their status vis-à-vis men, and can potentially transform gender relations. Agarwal (1994) insists that 'women's control over economic resources is crucially mediated by non-economic factors' (264). Such focus exemplifies the feminist environmentalism and the WCD approach: keeping sight of the 'economic' as a key conduit for women's inferiorization without privileging the economic above all quotidian aspects of women's lives. 'Culture as lived experience becomes a means for mapping how inequalities are confronted and remade, and thus relevant because culture offers a non-economic, yet material way to "produce knowledge" and design alternative strategies in struggles for social change' (Bhavnani et al. 2003, 9).

While a focus on gendered power and distribution of household resources remains a key focus of GAD approaches, such strategies can be limited in practice. GAD approaches disappoint in the way in which social differences, such as gender, are misunderstood and operationalized as 'metrics of technocratic and efficient decision making' rather than 'politically contested categories and outcomes of power' (Radcliffe 2015, 53). This neoliberal approach, as GAD scholars purport, dulls gender equality objectives and promotes 'an exclusively material understanding of empowerment . . . that remains underscored by patriarchal and gendered relations' (Wilson 2015, 807).

Feminist political ecologies

Informed by a variety of influences, including gendered environmental activism and GAD critiques, feminist political ecology has recently shifted to more critically feminist directions. A recent advance in the subfield, referred to as 'new feminist political ecologies' (Elmhirst 2011), is demonstrated by increased attention to the mutual embeddedness of multiple social differences. FPE is informed by the insights from post-colonial feminist critiques of development, particularly Chandra Mohanty's influential paper 'Under Western Eyes' (1986). In this work, she argues that 'Third World women' are represented in gender and development narratives as 'a homogenous "powerless" group often located as implicit victims of particular socioeconomic systems' (338). Such a presupposition assumes that all women in the Global South are victims and that their salvation lies in Western-driven assistance. These racialized representations are increasingly the source of creative development critique from feminist scholars looking to illuminate how such representations are naturalized, taken for granted, and thus serve as justification for coercive development interventions. While Mohanty and Black feminist scholarship is longstanding, and was taken up by feminist geographers prior to 'new feminist political ecologies' (Asher 2009; Gururani 2002; Mollett 2010; Radcliffe

Environmental/feminist struggles

and Pequeño 2010; Sundberg 2004), feminist political ecology has recently (as a field) illustrated a more consistent attention to Mohanty's insights, Black feminist thinking and GAD ideals, and more critical—intersectional—forms of gendered analysis in international development thinking.

Race, reproduction, 'nature,' and development

Population control is a growing focus in feminist political ecology (Sasser 2014). Feminist scholars link population control policies and practices to past eugenicist concepts of race (Sasser 2014). Similarly, Neo-Malthusian theories are closely related to population control. Malthusian ideas became prominent in the 18th century, where population growth was thought to be a central threat to global food production. This sentiment was re-ignited in the post-war development narratives of the 1940s and 1950s. In the context of decolonization in the Global South, Malthusian narratives presumed that there would be increased migration to the Global North, accompanied by food shortages. Such thinking is still firmly lodged in contemporary development thought. For feminist political economists and FPE scholars, population discourse is characterized by 'its reduction of "Third World women" to their reproductive organs and specifically their wombs, which are pathologized as "excessively reproductive" and requiring intervention' (Wilson 2015, 813).

Such gendered constructions of race continue to be imbued in population control policies in this neoliberal moment. For instance, women in the Global South are denied access to safe contraception, and are instead subject to coercive population control policies designed by development institutions in the Global North. Indeed, forcible sterilization remains common. According to feminist geographer Kalpana Wilson, during the period between 2003 and 2014, coercive sterilization resulted in 12 deaths per month in India (Wilson 2015, 814). Coercive sterilization remains a component of development practice throughout the Global South, mandated through Global North donor programs.

For feminist scholars, population control is underpinned by many myths. These include fears of food shortages that, despite technological innovation in food production, are circulated through climate change discourses, raising concerns that the Global North will become hosts to millions of hungry climate refugees. Population growth is also blamed for food crises associated with land grabbing by transnational and foreign governments who buy, and sometimes illegally control, land in foreign countries to produce food for their own domestic populations, often leaving local populations hungry and landless (*Economist* 2011). Under current mainstream development thinking, population control would 'mitigate' these crises and even lead to a decrease in maternal mortality rates and an increase in child survival rates—factors included in 'mainstreaming' gender equity and which remain longstanding development priorities. However, what is missing in this equation is that population control policies do not operate alone.

Indeed, gender equality is central to the United Nations' Sustainable Development Goals (SDGs), which aim 'to end poverty in all forms everywhere' (www.sustainabledevelopment.un.org). But Wilson (2015) contends that gender equity policies are meaningless because they are channeled through the ever-popular neoliberal policies, such as smart economics. While praised by mainstream development organizations, such as the World Bank, smart economics is a rationale that identifies a pathway to gender equality by maintaining that poverty reduction can be abated through investing in women and girls with 'intergenerational benefits' (Chant and Sweetman 2012). However, for critics, smart economics risks the entrenchment of women's naturalized responsibilities for both productive and reproductive labor and presupposes that women will always work harder, and produce more, than men. Hence, gender equality curiously translates to an increase in women's participation in labor markets, driven by a woman's imagined and seemingly innate gendered drive to spend a higher share of earnings on children. Such development imaginaries then 'instrumentalize' women's labor without acknowledging that these policies rely upon 'patriarchal structures and gendered relationships of power' (Wilson 2015, 807; Cornwall and Edwards 2014). Furthermore, these imaginaries resemble 'the moralistic overtones... of the Victorian narratives of the "deserving" and "undeserving" poor, and like them, are deeply racialized in their re-inscription of essentialized constructions of men in the Global South as inherently "lazy," irresponsible, and pre-occupied with sensual pleasure' (Wilson 2015, 811). Thus, it appears that gender equity policies, at least in the case of smart economics, are in fact built upon notions of racialized gender, and operationalized racialized gender stereotypes about 'Third World women' and 'Third World men.'

This corporate-led development model posits that adolescent girls are the way of the future (Figure 7.4): Nike's notion of the 'Girl Effect,' the UNs 'Girl Up,' and Plan International's 'Because I Am a Girl' are popular slogans that uncritically apply WID approaches, while ignoring the kinds of social conventions that limit young girls' empowerment through other types of power relations (Chant 2012; Wilson 2015; Sasser 2014).

Environmental/feminist struggles

Figure 7.4 Investing in girls: is it this easy? Bangladesh, Mithamoin, Kishorgonj. Students of Tomija Khatun Model girls' school with packs of sanitary towels they have been given. The teenage students are happy as previously they used cotton cloths which caused infections among many girls. Photograph by G.M.B. Akash. Reproduced courtesy of Panos Pictures. (www.panos.co.uk/preview/00227017.html?p=1)

Such practices do little to erase existing dominant economic models, gender inequalities, or racialized subjectivities, and only intensify present and future expectations of women's labor. The naturalization (taken-for-grantedness) of women's exploitation, masked as 'empowerment,' does not simply operate in the context of employment and population control, but unfolds in and through environment-development interventions as well.

Biodiversity conservation and land struggles

FPE emphasizes gender relations as an important shaper of natural resource conflicts and environmental transformation. This multi-scaled approach simultaneously attends to global environmental policy, national development programs, households, and the body. Once again, feminist political ecologists critically interrogate the way in which gender informs conflicts over nature, through a number of situated struggles and contexts including land titling, commercialized agriculture, resource extraction, urban restructuring, biodiversity conservation, and food insecurity (to name a few), and the ways in which these developments challenge and empower men and women differently. FPE strives for a more complete picture of people's lives in the Global South, beyond economic indicators and the so-called 'win-win' scenarios of global conservation policies at the community levels. FPE demands attention to the strategies of resistance to what is now the ubiquity of neoliberal development practices shaping nature–society relations in old and new ways.

As a feminist political ecologist, I research the politics of natural resource access inside the Honduran Rio Plátano Biosphere Reserve (RPBR). Honduras is a small country in Central America. Through my research, I show that while global initiatives for protected areas attempt to 'humanize' protected area space, for the purpose of sustainable development, and move away from fortress models of conservation that restrict human activity, biodiversity conservation practices often **dehumanize** those living inside parks and biosphere reserves. While feminist political ecologists are staunch defenders of the environment and care deeply for nature as we work to mitigate climate change, at the same time, we are deeply concerned that global initiatives have come to legitimate the erosion of environmental justice for people, wildlife, and communities living inside protected areas and other environmental enclosures (Walsh 2015; Mollett 2015). For instance, the management of the Rio Plátano Biosphere Reserve, while based in Honduras, is influenced by global environmental policies and programs such as REDD+, Rio +20, UNESCO's Man and the Biosphere Program, and the International Union for Conservation (IUCN). Such agencies and bilateral funding (nation–nation) provide financial and technical assistance on how best to 'protect' biodiversity.

In Honduras, the RPBR is home to indigenous Miskito, Pech, Tawahka and Afro-indigenous Garifuna communities. They are also joined by Creole and *ladino* communities and together number more than 20,000 people living inside the RPBR's cultural zone (Figure 7.5). Indigenous, Garifuna, and Creole communities have historically occupied the coastal regions of the Mosquitia rainforest (the name of the region where the RPBR is located), since before Honduras became a republic in 1815.

In 2002, the Honduran state, in the name of **sustainable development** and **biodiversity protection**, and with support from the German government, proposed a land-titling project called Catastro y Regularizacion (Mollett

Environmental/feminist struggles 7 Chapter

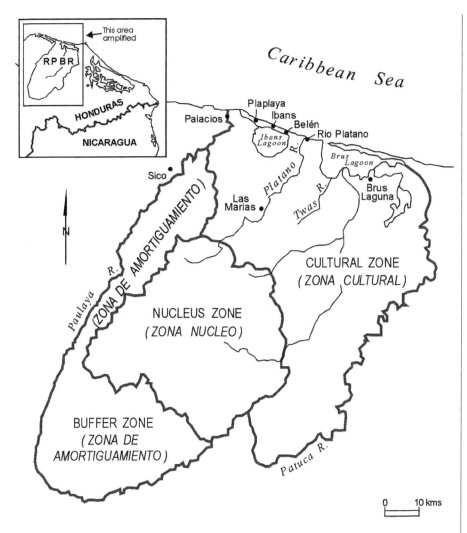

Figure 7.5 Rio Plátano Biosphere Reserve map. Permission from Latin American Research Review (LARR); Mollett 2006.

2010, 2011). This land registration program originally required that indigenous communities individuate their customary family lands. For many of these communities, land is held collectively and shared among family members without ownership titles. These arrangements link present communities to their ancestors and are imbued with cosmologies that treat non-human nature not simply as economic resources, but as spiritual relatives and neighbors. However, under the project, and in the name of biodiversity conservation, project officials proposed privatizing these lands by offering use rights to families, in the name of one person (the male family head). This state demand, while eventually canceled more than ten years later, set off a number of struggles between, and among, community members. In the Miskito community of Belen, a number of factors sparked tensions between Miskito men and women, both within households and across generations. Such factors were:

- the move from collective ownership to the individual holdings;
- state claims that failing to report land holdings would formally revert 'unclaimed' lands to 'state' territory;
- the limitation of rights to use rights only and not ownership rights.

These proposed changes stirred great debates among community leaders and the project officials, and have shaped Miskito women's access to family lands (Mollett 2010). As FPE scholarship asserts, in many cases, formalized land tenure programs often privilege males as heads of households and exclude women (presumed to be dependents) from ownership opportunities. This occurs despite the fact that, in Miskito communities, women carry a disproportionate share

Figure 7.6 A Miskito woman farmer. In the RPBR, Miskito women participate in farming, often on family land on the coasts and upriver. In this plot grow, yuca, malanga, and lemongrass, among other crops. Miskito peoples also grow fruit trees. While the fruit is important to Miskito diets, the trees denote the boundaries (however porous) of family lands. In 2014, Miskito Federations inside the RPBR received collective titles to lands after many years of struggle. This has true significance for all Miskito peoples, but especially Miskito women who would have been the first to lose land control had their lands been individuated under a previous state proposal. Reproduced courtesy of Sharlene Mollett.

of household and community responsibilities and social reproduction. These ongoing processes of women's exclusion necessitate increased interrogation of who benefits from development financed titling and registration projects. While FPE exposes the gendered inequalities within our research sites, it is important to note that not all women openly problematize gender asymmetries, particularly when their communities are also exposed to racialized power inequalities. Rather, at times, women actively look for ways to engage strategically with the local forms of patriarchy, as a way to strengthen community resistance to state racism in land distribution (Mollett 2010).

For instance, Miskito women quietly acknowledge that the imperative to register land individually threatens to favor men at the expense of the solidarity of family lands, and despite the fact that Miskito inheritance is matrilineal. Under the Catastro y Regularizacion project, women's participation in the multiple sites of production and longstanding relations to natural resources through cultivation, worship, housework, water procuration, cooking, gathering, etc., are completely ignored by privatization.

While ownership is largely sought as a means of status and livelihood security, access to land in male-dominant societies can protect against various insecurities for women. Land provides direct benefits through food cultivation and tree growing and (in some contexts) can serve as credit for investment in farming. Additionally, land is a transferable asset during times of crisis. Land can improve the chances of finding, and the quality and conditions of, wage labor. Thus, it is an important asset for the reproduction of rural off-farm work, because many perceive that 'having land' makes one a more 'stable' and 'desirable' employee (Agarwal 2003).

Most recently, the Miskito Federations inside the RPBR received collective titles to their ancestral lands. These collective titles are held by Federations and Territorial Councils. While this is a historical victory for the Miskito people, and cannot be understated, the ways in which gendered struggles emerged over the threats of privatization remain salient to gender and land debates in feminist political ecology, and raise different questions over the kinds of land control that collective titles offer to indigenous, Afro-descendant, and *ladino* residents inside the RPBR, and how global biodiversity conservation policies actively produce conflict at local scales.

Land grabbing

Land continues to dominate development debates, particularly with the subject of **land grabbing**. Land grabbing commonly refers to the 'contemporary' emergence of transnational, large-scale, and elite land transactions. These are both commercial (i.e., industrial agriculture) and personal (i.e., residential tourism). Such land transfers often facilitate much-needed food production, with the support of international development agencies, but are also conducted for the purpose of biofuel, timber, and mineral extractions, which often end in large-scale displacement from land for local people (Alonso-Fradejas 2012; Oxfam International 2016). While present-day land grabs are in large part shaped by the legacies of colonial land control, researchers tend to focus on the way in which agrarian landscapes are full of novel actors, crops, laws, and labor strategies that disrupt land and natural resource ties for many rural and urban poor. For instance, in addition to former colonial powers now located in the Global North, land grab actors are joined by governments and companies from the Global South, such as east and southeast Asia, the Gulf States, Brazil, and China. These actors are capitalizing on their development success in relation to weaker economic states, who often allow them to appropriate land through foreign direct investment projects, often to secure food security and satisfy energy needs in their own countries. These projects, while sometimes employing local people, raise ethical questions about the exploitation of local impoverished communities and the export of food provisions out of a country. These groups tend to remain landless for the purposes of feeding foreign populations located outside national borders (i.e., Chinese food production in Africa) rather than within their own countries (Zoomers 2010).

In response to the focus on contemporary land grabbing, understood through largely rich/poor and foreign/domestic binaries, feminist political ecologists highlight a more historically and spatially informed process of land grabbing that occurs over time. FPE shows how land grabbing (and land conflicts alike) is a product of older power structures and inequalities that are linked to place and space and the powers that shape such geographies, such as race and gender (Mollett 2016; Verma 2014).

There is remarkable silence regarding the way in which the global land grab is intensely gendered. The mainstream literature on global land grabbing often focuses on 'acquisitions' and 'investments.' This approach views land grabbing through dominant development discourses that focus on privatization, commoditization, enclosure, land markets, resource extraction, external food security, and biofuel production. Imbued in these narratives is neoliberal

Environmental/feminist struggles

economic thought that assumes that land functions solely as a resource for profit 'through the unwavering subscription to a "free" market economy and ethos of privatization' (Mackenzie 2010, 37). This practice reduces the meaning of land to its 'technocratic and productive' values and erases the ways in which land has socio-cultural, political, historical, and gendered meanings (Verma 2014).

Thus, not all land grabs are in fact large-scale acquisitions, but rather they are a culmination of years of land loss experience through everyday gendered and racialized processes (see also the case of Bahia, Brazil, below) (Verma 2014; Mollett 2016). Communities are undoubtedly at times forcibly removed from space for large-scale land deals. But more commonly, forms of dispossession occur through micro-political and more mundane forms of land grabbing over extensive periods of time. Often such micro-grabs then facilitate more spectacular land-grabbing events.

Figure 7.7 Land grabbing in Myanmar. Myanmar (Burma), Hpa an. Two women stand outside a rural house near Hpa an. One of the women wears a thanaka face mask, a powder made from ground bark used for a sun block and a skin treatment. Farmland in the region is under threat from land grabs by the growing concrete production industry. Photograph by Patrick Brown. Reproduced courtesy of Panos Pictures. (www.panos.co.uk/preview/00225184.html?p=16)

The nuance and complexities highlighted in the field of feminist political ecology are important contributions to international development policy. Feminist political ecologists continue to investigate how proposed development policies claim to ameliorate the structural inequalities of poverty vis-à-vis women's land control and participation in rural communities. For instance, recent research with smallholder farmers and rural communities in Mexico and Peru suggests that conditional cash transfers (CCTs), while dubbed as 'revolutionary' forms of poverty alleviation, have mixed results for rural women and that the conditionalities tied to these protections re-entrench gender inequalities and may undermine women's land control, despite extra income (Radel et al. 2016; Cookson 2016).

Land, water, and spirituality

For many people, land and natural resources are infused with deep spiritual and cultural values. Land and natural resources root identity, and link present communities to their ancestors and to culture and traditions. These connections, particularly for indigenous and Afro-descendant communities, offer authentication of their place in the nation-state and/or as a nation, i.e., First Nation Peoples and Native Peoples in North America. Feminist political ecologists are attentive to how natural resource struggles are not simply economic struggles. Our work increasingly pays attention to 'more-than-human-natures' and the will to disrupt ontological hegemonic thinking that places nature outside of culture and separate from human beings (Sundberg 2014; Hovorka 2012). These constructed divisions have legitimated human intervention in and control of, and justified the appropriation and exploitation of, nature (i.e., mining, logging, and other forms of extraction).

In April 2016, the Standing Rock Sioux tribe began a protest at the Sacred Stone camp near Cannonball, North Dakota, to oppose the planned Dakota Access Pipeline (DAPL). Since then, the camp has grown, as activists from over 120 Native American 'Nations' or tribes have come together and stood in the path of the pipeline, stopping its progress (Figure 7.8).

Accordingly, particularly in the case of Latin America, there is growing consciousness among development critics that recognizes how

> Nature, of which humans—living and dead—are a part, and the cosmos together configured, reproduced and ordered life, and with it modes of

Environmental/feminist struggles

Figure 7.8 In defense of human rights. United States of America, Cannonball, North Dakota. A woman astride a horse in the Sacred Stone protest camp. The proposed 1,172-mile pipeline would cost 3.5 billion USD and carry about 470,000 barrels of crude oil a day. The pipeline's route would take it under the Missouri River, at a point just upstream from the Standing Rock reservation which, Sioux tribal members say, would mean their drinking water was vulnerable to pollution and their sacred sites threatened. Photograph by Hossein Fatemi. Reproduced courtesy of Panos Pictures. (www.panos.co.uk/preview/00223127.html?p=1)

gender grounded in the practical everyday activity of women and men, and in the fluidity between feminine and masculine forces, forces that superseded the anatomical body. Colonial-imperial domination from the Incas to the Spanish and beyond worked to undo the fluidity and ties. At the same time, it worked to install hierarchies, dichotomies and divisions based on the "ideas" of race, gender, sexuality and nature and on the social systems of hetero-normativity, patriarchy and racialization. The control, domination, exploitation and taming of the uncivilized, savage, irrational and wild, i.e. nature, 'Indios', 'blacks' and women (and most especially indigenous and African women), thus became 'naturalized' mechanisms and necessities of modern/colonial order.

(Walsh 2015, 104)

The **coloniality of nature** reifies a singular model of nature as civilization that simultaneously constructs, justifies, and legitimizes power hierarchies and structures across the world, so that patriarchy, modernist development paradigms, and coloniality appear normal (Walsh 2015).

Indeed, the woven links between human and non-human natures enrich the field of feminist political ecology in spiritual ways. In the seaside town of Gamboa de Baixo, on the coast of the Bay of All Saints in the city center of Salvador, Bahia, Brazil, a female-led grassroots movement leads resistance against land displacement and access to the sea on behalf of long-time residents. In this Brazilian fishing community, as Keisha-Khan Perry's (2009, 2013) research describes, Gamboa de Baixo struggles occur in the context of threats of mass evictions due to urban renewal projects, or what activists call 'a wave of black clearance' (2009, 10). This is not simply a collective defense against land expulsion, but what drives this movement is their spiritual ties to, and love of, the sea that lines their coastal community. This movement is underpinned by the community's belief in the African diasporic religion of Candomble. Within such is Iemanja (Yemanja), an exalted goddess of the sea, referred to as the 'mother of the waters' (Perry 2009). Annually, February 2 in Salvador marks one of the significant celebrations of Iemanja. The largest celebration takes place in the Rio Vermelho neighborhood (a predominantly white neighborhood)—a festival that is both community celebration and a key attraction of national and international tourism.

Historically, however, this traditional celebration is practiced by Black urban fishing communities in less spectacular fashion in the community of Gamboa de Baixo. This cultural expression is marked by local people paying homage and offering prayers to the goddess of the sea in hopes that she will protect fishers and their livelihoods, upon which the local economy and the Afro Brazilian culinary traditions rely, and for the protection of their children as the future inheritors of coastal lands and community culture (Figure 7.9).

For activists, these festivals have wider and more diasporic meanings. African religious traditions and their worship of the sea is a way in which Afro Brazilian women make claims to urban land on the Bahia coast (Perry 2013). In terreiros (Candomble houses), Afro Brazilian women practice African religious traditions, passed down from mothers, grandmothers, aunts etc. But also, the past endows them with rights to the land upon which they practice these traditions. Indeed, 'black women have been uniquely positioned in these communities as having both collective memory and legal documentation of ancestral lands' (Perry 2009, 11). Such memory reaches beyond the Bay of All Saints and across the Atlantic to African women as the customary land owners, with matriarchal and matrilineal responsibilities. These are not just spatial performances,

Environmental/feminist struggles

7 Chapter

Figure 7.9 Yemenja celebrations. Brazil, Salvador, Bahia. Worshippers honour Orisha Yemenja on the evening of Day of Yemenja, or *Dia de Yemenja*. Orisha Yemenja is the goddess of the sea and her day is celebrated by thousands of Brazilians at Praia Velmelho. Devotees offer gifts such as flowers, objects of female vanity, Yemenja statues, and sweets to the goddess and leave them in the water.

Yemenja is an Orisha, a saint that represents the natural world. She is viewed as the essence of motherhood, the protector of children, fishermen and sailors, and most importantly, she is the sea itself. She wears a dress with seven skirts that represent the seven seas. Originally Yemanja was a river goddess of the Yoruba in Nigeria. African slaves brought Yemenja to the New World, where she became the Goddess of the Ocean. Photograph by Lianne Milton. Reproduced courtesy of Panos Pictures. (www.panos.co.uk/preview/00216983.html?p=10)

but in fact, contemporaneously, urban land in Brazil 'represents the ability for Afro Brazilian women to pass spiritual and material resources from one generation to the next. Land has become one of the greatest social and cultural assets for black people and particularly for women, who are the most economically marginalized' (Perry 2009, 11). This struggle for land, across the Atlantic and linking past and present, couples demands for collective property rights and the preservation of material and cultural resources. Such demands are political. Collective resistance against the violence of land expulsions exemplifies Afro-descendant 'sensibilities of gendered racial liberation and social transformation' in urban Brazil (Perry 2009, 11).

Resistance to displacement in Gamboa de Baixo is not simply about urban land and access to the sea. The struggle is also a contest over the right to clean water and sanitation. These struggles, in Brazil and elsewhere, make clear that environmental justice includes the ability for residents to flush toilets so that feces actually disappears, without which disease (such as cholera) is imminent. Demands for clean water (for drinking and sanitation) and a more 'dignified life' represent Afro Brazilian demands to participate in plans for urban renewal, a process that currently excludes their voices. But they do not sit quietly by, waiting for change: local activists organize garbage removal from the ocean and have adopted biodegradable materials in their religious celebrations (Perry 2013). Drawing insight from feminist geographer Katherine McKittrick, Perry insists that 'black urban spaces are racialized and gendered "terrains of domination," where black women's politics are deeply connected to resistance against "geographic domination" as practiced in environmental neglect, violence and displacement in Brazilian cities' (Perry 2009, 13; 2013). The case of Gamboa de Baixo suggests important directions in feminist political ecological understandings of struggles over environmental change; they are simultaneously deeply racialized and emotional, and require a broad and sincere acknowledgement that such struggles cannot be explained through gendered inequalities alone.

Emotional political ecologies: bodies, water, and embodied subjectivities

In Delhi, India, city residents rely on a variety of (frequently illegal) sources and practices as a way to access water and sanitation. As water access is unreliable, even within elite neighborhoods in Delhi, the combination of unequal distribution and poor-quality water systems leaves most household water and sewage requirements unfulfilled (for an exception, see Figure 7.10). While most gender and development scholarship emphasizes how water procurement is a highly gendered practice, it is also important to note that it is highly **emotional work**. There are embodied consequences of water inequality, shaped around processes of social and spatial differentiation and the shaping of quotidian lives in the city.

Water inequality in Delhi is partially influenced by a history of planning that brought in migrant laborers to make Delhi more orderly for its residents, but without a vision or a place for the people that actually labored to remake the city. Baviskar (2003) notes that 'the building of planned Delhi was mirrored in the simultaneous mushrooming of the unplanned Delhi' (91). This resulted in migrants and poor workers, who were once required for urban

Environmental/feminist struggles

7 Chapter

Figure 7.10 Water procurement in New Delhi, India. Residents gather at a Delhi Jal Board water supply tanker to collect water in the slums of Govind Puri. The Delhi Jal Board distributes multiple tankers of free water to all residents in the slum area. Photograph by Sanjit Das. Reproduced courtesy of Panos Pictures. (www.panos.co.uk/preview/00212320.html?p=4)

development, to be pushed to live on the city's periphery—ironically the area identified for demolition and renewal. In fact, the informal economies and mixed land uses of slums were so salient to the urban renewal efforts of the Delhi government that the urban poor were often extended amenities by city officials. However, since the mid-1980s, the deindustrializing of the city and the limiting of formal employment, which was accompanied by the incursion of global circuits of finance and services, have been linked with a persistent criminalization of the poor, a process that leaves questions about the meanings and practices of rights and urban citizenship (Baviskar 2003; Doshi 2017; Truelove 2011).

Still, Delhi residents are quite resourceful procuring water and accessing sanitation, however crude, from discovering open taps and water supply tankers, to accessing quasi-legal water connections, urban ponds, and hand pumps and walking far distances to find a place to defecate. Because the price of piped water is exorbitant, despite state subsidies, the costs to the poor are disproportionate since they mainly access water from commercial non-state sources. As the responsibility to source water is practiced by women and girls, the concomitant precarity of procurement, associated with both water access and sanitation, is saliently differentiated. Research in Delhi shows that 'women's bodies encounter differing degrees of gendered hardships, physical labor, and public shame that are shaped by their situated position within families, communities, and class groups in the city' (Truelove 2011, 147).

Embodied consequences of a lack of access to water and sanitation highlight a number of challenges:

- the harsh labor and physicality of water procurement;
- health issues related to contaminated water;
- physical abuse as a result of the criminalization of procuring water from illegal sources.

These embodied experiences, and others like them, disclose gender, caste, and class differences, and the physical constraints and economic hardships.

As a result, some women attempt to access water from their employers, particularly while working as domestic workers in elite and upper-class homes. These favors granted in access to drinking water require a disproportionate amount of energy to repay employers through extra work. As Truelove argues, 'the loss of a degree of control over their labor and negotiating power, coupled with the physical and emotional stress of sometimes working extra hours for less pay, indicates how the space of the work place takes on new gendered meanings and constraints' (Truelove 2011, 148).

There are also embodied consequences to accessing sanitation. When women have no access to sanitation, or latrines are not in proximity to their homes, they risk rape, harassment, kidnapping, and ridicule as they search for a place to urinate and defecate. Staying close to home is also a source of great anxiety for women. Frequent stomach illnesses and women's menstruation during the night mean female residents are forced to go directly outside their dwellings within earshot of their neighbors. Female residents experience physical and emotional ridicule 'where their bodies are caught in the nexus of local cultural relations (which ascribe a sense of shame to the visibility of women's sanitation practices)' (Truelove 2011, 148).

As in Delhi, the physical and emotional consequence of unequal access to water is also illustrated in the context of Bangladesh. According to feminist political ecologist Farhana Sultana (2009, 2011), men and women articulate how resource access and conflicts are deeply emotional and invoked through different meanings of 'suffering.' A focus on suffering advances development

Environmental/feminist struggles

research by interrogating the material and emotional hardships of water procurement and consumption. This approach counters a pattern in development research that ignores subjectivities. Feminist political ecology pays increasingly close attention to emotions as a way to make visible women's agency. In Bangladesh, a lack of safe water due to contaminated tube wells shapes people biologically, but also informs their claims to water. 'Suffering for water' as well as 'suffering from water' leads some to employ cultural and religious (moral) obligations to share water, while others articulate poverty and sufferings to produce sympathy as a way to make claims (Sultana 2011, 163).

For feminist political ecologists, while it is important to highlight divisions of labor and patriarchal relations of power, it is also important to recognize that wives, mothers, daughters, sisters, and daughters-in-law do not just follow patriarchal rules of labor, but gendered identities, behavior, and self-worth are constituted in, and through, their various environments, shaped by caste, race, rurality, indigeneity, age, and the like (Nightingale 2011; Gururani 2002; Mollett 2010; Mollett and Faria 2013). In the Kumaon forests of the Himalayas, forests as public and shared spaces not only serve as the places where livelihoods are made, but are also key 'cultural spaces' that are comprised of women's work experiences and memories of labor that invoke gendered subjectivities in temporal, cultural, and emotional ways. Forests illustrate the interwoven relations and subjectivities that 'shape gendered relations in nature and illustrate how marginal gendered subjects are produced in nature through relationships of *pain and pleasure*' (Gururani 2002, 233).

Feminist geography brings a focus on **emotions** in political ecology that enriches a complex set of knowledges about gender, difference, development, and nature, revealing the ways in which people and their communities experience poverty, struggle, and inequality, and how subjectivities are shaped through struggle and compromise. Emotions are relational and they occupy the spaces between people, nature, and places. A focus on emotions, advanced through encounters, helps us to understand the material conditions of everyday life and enables us to gain heightened awareness of the politics of natural resource struggles and the making and remaking of environmental knowledges (Wright 2012). But, of course, tracking emotions is not only a way to advance understandings of hardship, but also resistance. In fact, 'struggle and suffering are not only situated in place but actively produce it' (Wright 2012, 1118). Thus, emotions are important for understanding the inner workings of the spaces of environmental social movements, and resistance against development-induced inequalities.

Figure 7.11 US veterans join Standing Rock protest. United States of America, Cannonball, North Dakota. US military veterans march with the 'Water Protectors' through the snow at the Oceti Sakowin Camp, one of several camps housing protestors fighting the planned Dakota Access Pipeline (DAPL). Photograph by Hossein Fatemi. Reproduced courtesy of Panos Pictures. (www.panos.co.uk/preview/00226635.html?p=3)

Conclusion

Men and masculinity

While most of the examples in this chapter reflect the tendency in FPE to highlight women's gendered subjectivities, important work in FPE also builds upon intersectional thinking and brings necessary attention to the way in which men and masculinities, in addition to women and femininities, are specifically embroiled in environmental struggles. Indeed, part of 'new feminist political ecologies' is an effort to take on gender, with more attention to men and masculinities (Nelson 2012; Wangui 2014). For instance, in her study 'A Feminist Political Ecology of Livelihoods and Intervention in the Miombo Woodlands of Zambézia, Mozambique,' Ingrid Nelson (2012) demonstrates the ways in which the practices and enactments of 'manhood' are shaped in

Environmental/feminist struggles

relation to timber logging and a history of livelihood relations in Mozambique. In her research, Nelson shows how men perform masculinity through singing songs while hauling timber. This collectively learned practice takes place throughout Zambézia. The content of the songs emerges from men's lives and draws inspiration from love and household tensions and responsibilities, disdain for the boss, struggles for survival, and reflections on the country's colonial past. Such performances also influence the power of employers and development practitioners to accumulate timber and execute the goals of a particular project. Nelson's important contribution maintains that:

> If we see gender identity as a process tied to ecology, economy and historical context, and if we step away from equating gender with women or relations between men and women, then we can develop a much deeper analysis of how loggers succeed or fail in extracting timber and what environmentalists [and land-grabbing scholars alike] are missing in the socio-ecological dynamics.
>
> (165–7)

While Nelson's project makes clear the need to attend to masculinities, this work also significantly represents how, in spite of salient challenges, including poverty and violence, forms of resistance map spaces of future possibilities and just forms of change. Attention to resistance highlights that despite the violence of gendered exclusion in environmental struggles, femininities and masculinities (not as binaries but as relations) are part of the spatialization of resilient subjects in Zambézia and thus we need to chart both in our work in FPE.

Resistance and climate justice

Feminist geography links more spectacular resistance movements, such as Women Human Rights Defenders and the Chipko Movement, to intimate, quiet, and subtle resilience–resistance movements that seek not only to secure economic survival, but also to 'recuperate dignity' lost through oppression (Katz 2004; IM-Defensoras 2015). By resilience, I am referring to feminist geographer Cindi Katz, who through her research on globalization and agricultural change in Howa Sudan helps us to understand resilience as a 'reformulation of the very grounds and compass of everyday life' (Katz 2004, 245). Such dignity has been equated with what James Scott refers to as the 'minimal cultural decencies that serve to define what full citizenship in local society means,' which allows people to survive and sanctify collective resources and knowledges as a way to 'rework' and 'resist' forms of domination that shape their lives (Scott 1985, cited in Katz 2004, 246). In a similar way, Laura Shillington (2013) demonstrates how residents make their claims to the 'right to the city' through urban agricultural practices. In San Augusto, Nicaragua, patios and fruit trees are the products of 'home ecologies' made 'through the complex relations within and between home, community, city and globe' (103). Such patio ecologies are, however, not necessarily aligned with the vision of urban planners and development practitioners who wield other forms of criteria for abating poverty. In this case, 'home is an important space where such competing ideals are played out' (103).

Feminist geographic focus on the 'everyday' attends to multiple scales and furthermore contributes to important understandings about climate change. It is true that political power influences environmental change through international summits like the Paris Convention, but the daily decisions made by government officials, state employees, and community residents shape climate governance (Bee et al. 2015). Within this kind of governance, policies are increasingly informed by neoliberal reasoning that turns 'green' initiatives into markets without considering how this approach has profoundly inequitable impacts on people. As Bee et al. (2015) write, '[p]erhaps the most problematic contradiction of neoliberal climate governance is that the focus on individual action in neoliberal climate governance deemphasizes the wider political economic context under which climate change is produced' (7)!

Taking heed of the everyday, as a way to understand equity in climate change governance, also makes visible disparate environmental knowledges. Until recently, climate change research has commonly contextualized discussions of women through their 'vulnerability' rather than highlighting women's agency and environmental knowledge. FPE scholars argue that such knowledge of environmental change remains an important source of potential climate change innovation. Climate change innovations are important for the design of future climate adaption strategies, and understanding natural resource scarcities and climate-related livelihood challenges. For instance, Stephanie Buechler's (2015) work demonstrates how dairy producers and women cheese makers employ 'creative tactics' in response to growing climate-related water scarcity in Sonora, Mexico (99). In a similar way, Dina Najjar (2015) describes the way in which women in two desert resettlements, as part of the Mubarak Resettlement Scheme in Egypt, are adjusting to new landscapes and property

Environmental/feminist struggles

Figure 7.12 Resilience in the face of climate change in Kenya. A Maasai boy, cattle, and sheep at a check dam in Magadi, Kenya. The Maasai call this rainwater-harvesting structure *osilange*. It is made out of an earth dam in an area where rainwater flows down. This is a traditional method of harvesting rainwater and a good example of how to adapt to climate change and global warming. After a good rain season, this *osilange* can provide water to people and animals for six months. The water is not only used for drinking by animals but also by the Maasai, who boil it at home or drink it from its source without filtering. Photograph by Dieter Telemans. Reproduced courtesy of Panos Pictures. (www.panos.co.uk/preview/00166787.html?p=37)

regimes. In this case, Najjar focuses on how women work to maintain access to water through the implementation of water conservation activities and keen environmental knowledge of planting crops with minimal water requirements, without assistance from the state.

The way in which FPE scholarship highlights both marginalization and resistance is important for understanding both political ecological critiques of development, and a more global equitable distribution of natural resources. The colorful and ethnographic detail provided by these FPE scholars helps illuminate intersectional gendered struggles for land, reproductive rights, water, forests, and food. The very dynamics that shape struggles over land and water inequalities are those that inform violence against women, women's employment discrimination, and other legal geographies of marginalization (i.e., property rights).

While in this chapter I have primarily focused on the Global South and development contexts, the spaces of environmental inequalities are global, share both global and local forms of power, and require multi-scaled responses. In this way, the work in feminist political ecology shares motivations with geographers working to understand and resist environmental racism in the US. Thanks to the work of Laura Pulido, we can no longer reduce environmental injustices to class inequalities. But rather, various racisms (white privilege, settler colonialism, white supremacy) blend with myriad forms of gender and class power that produce the toxic spaces where a disproportionate number of people of color are compelled to live, or try to live, in the face of polluted air, water, land, and the like (Pulido 2016). For Pulido, 'the time is ripe for a deep engagement with racial capitalism' (Pulido 2016, 3)—a position that many feminist political ecologists, and feminist geographers, increasingly embrace.

Recommended reading

Buechler, Stephanie, and Anne-Marie Hanson. 2015. *A Political Ecology of Women, Water and Global Environmental Change.* New York: Routledge.

Doshi, Sapana. 2017. 'Embodied Urban Political Ecology: Five Propositions.' *Area* 49 (1): 125–8. doi:10.1111/area.12293.

Elmhirst, Rebecca. 2011. 'Introducing New Feminist Political Ecologies.' *Geoforum* 42 (2): 129–32. doi:10.1016/j.geoforum.2011.01.006.

Hawkins, Roberta, and Diana Ojeda. 2011. 'Gender and Environment: Critical Tradition and New Challenges.' *Environment and Planning D: Society and Space* 29 (2): 237–53.

McKittrick, Katherine. 2006. *Demonic Grounds: Black Women and the Cartographies of Struggle.* Minneapolis, MN: University of Minnesota Press.

Mollett, Sharlene. 2010. 'Esta Listo (Are You Ready)? Gender, Race and Land Registration in the Rio Plátano Biosphere Reserve.' *Gender, Place & Culture* 17 (3): 357–75.

Mollett, Sharlene, and Caroline Faria. 2013. 'Messing with Gender in Feminist Political Ecology.' *Geoforum* 45: 116–25. doi:10.1016/j.geoforum.2012.10.009.

Perry, Keisha-Khan Y. 2013. *Black Women against the Land Grab: The Fight for Racial Justice in Brazil.* Minneapolis, MN: University of Minnesota Press.

Rocheleau, Dianne E., Barbara P. Thomas-Slayter, and Esther Wangari, eds. 1996. *Feminist Political Ecology: Global Issues and Local Experiences.* New York: Routledge.

Sultana, Farhana. 2011. 'Suffering for Water, Suffering from Water: Emotional Geographies of Resource Access, Control and Conflict.' *Geoforum* 42 (2): 163–72. doi:10.1016/j.geoforum.2010.12.002.

Sundberg, Juanita. 2014. 'Decolonizing Posthumanist Geographies.' *Cultural Geographies* 21 (1): 33–47. doi:10.1177/1474474013486067.

Walsh, Catherine. 2015. 'Life, Nature and Gender Otherwise: Feminist Reflections and Provocations from the Andes.' In *Practising Feminist Political Ecologies: Moving beyond the 'Green Economy,'* edited by Wendy Harcourt and Ingrid Nelson, 101–28. London: Zed Books.

Wilson, Kalpana. 2015. 'Towards a Radical Re-Appropriation: Gender, Development and Neoliberal Feminism.' *Development and Change* 46 (4): 803–32. doi:10.1111/dech.12176.

Chapter 8

FEMINIST SPACES
CONCLUSION AND REFLECTIONS

Sharlene Mollett, Jennifer L. Fluri,
Risa Whitson, and Ann M. Oberhauser

> Genuine understanding must be rooted in a rejection of dominator culture . . .
>
> (bell hooks 2013, 57)

Conclusion

This book addresses the comprehensive scope and dynamism of feminist geography. We encourage readers to engage with the wide array of topics that make up this field. The themes and narratives in this book underscore our commitment to social justice and solidarity among not just women, but **all feminists**, regardless of gender, working for progressive and just change around the globe.

As such, **Chapter 2** focuses on the body as a site upon which social, cultural, and political meanings are inscribed, understood, and challenged. The sociocultural and political conceptualizations of gender are discussed as performative, meaning that gender is performed onto the body by repetitive acts of doing, rather than being a simple byproduct of one's biology. This chapter identifies the various ways in which social, political, and economic meanings are mapped onto different corporeal markers (such as gender, race, class, sexuality, and ability). This examination of corporeal geographies traces the gendered body, as it is understood in societies through sex and sexuality, as well as other intersectional identities. Corporeal identities are often linked with individual or collective experiences of being included or excluded from places, rights, or opportunities. Therefore, identifying or categorizing bodies based on one's gender, race, class, sexuality, or ability has been—and continues to be—a dangerous form of social control. These identifications and categorizations have been—and continue to be—used to justify enacting violence, killing, containment, and other forms of oppression on certain bodies. Making social, political, or economic assessments of one's character based largely on one's appearance further limits our ability to engage as equals with others because it does not take into consideration multiple and complex aspects of our identities.

In similar and different ways, **Chapters 3** and **4** explore the way in which our everyday spaces are imbued with gendered meaning. Such activities as walking down the street, riding the bus to work, decorating our homes, going shopping, going to a baseball game, taking a weekend getaway, and participating in an online forum are all moments in which our identities as gendered beings are being produced. During these moments, we are also

Conclusion and reflections

Chapter 8

constantly working to reproduce, maintain, or contest gendered norms and meanings. The spaces in which these activities occur—streets, parks, homes, malls, sporting venues, tourist attractions, and online spaces—are not simply neutral backdrops for the production of gendered meaning. Rather, these spaces themselves are saturated with gendered meanings, and it is through the negotiation between self, society, and space that our gendered identities and the meanings of gendered places are constantly recreated and renegotiated. The goal of these chapters is thus to highlight the way in which our ordinary, unremarkable, and familiar everyday spaces are complexly intertwined with our gender systems.

These chapters also emphasize the ways in which these relationships are fundamentally connected to the structural organization of our society. In this way, while gender is the central theme of this book, these chapters reinforce an important argument made both by feminist geographers and others: gender is only one element of struggles over power and belonging. As a result, these chapters support a central argument of this book: namely that the scope and charge of feminist geography is no longer to focus on women, as it was in the 1970s and 1980s, or even to give primacy to gender and sexuality, as in the 1990s and 2000s. Rather, as feminist geographers, we have the tools and frameworks available to us to explore and examine all of the ways in which power is exerted and negotiated in, and through, space and place to affect what we think, what we feel, who we are, and what we can become.

Furthermore, **Chapter 5** develops the economic dimensions of these social identities that influence how people engage in livelihoods to support themselves, their families, and their communities. These livelihoods are often embedded in unequal power relations that inform access to resources, and consequently opportunities, in the workplace. Feminist geographers have long analyzed and challenged these power relations and economic inequities. Understanding the connections between home and work, for instance, sheds light on the fluid and contested boundaries that shape gender inequality in occupations, pay, and promotions. Feminists' increasing attention to care work is also instrumental in understanding the often unrecognized aspect of many women's participation in the labor force. Economic globalization provides a critical landscape for feminist scholarship and praxis. Feminist economic geography examines the power relations and disparities that underlie neoliberal global capitalism. Labor migration, human trafficking, low-wage jobs in export-processing zones of South Asia, and the consumption of produce grown in the Global South are part of the circuits of capital and labor that are connected through gendered power relations. Finally, Chapter 5 addresses resistance and transgressive practices that are widely evident in grassroots organizing, diverse economies, and global networks that challenge the adverse effects of global capitalism.

Chapter 6 extends our understandings of global capitalism through body politics within the field of feminist political geography. While the body is not the only lens through which politics is examined, it has been a central component of analyzing gender and geopolitics. Gender intersectionality is examined by exploring the ways in which governments control citizens' access to sex, sexuality, and biological reproduction. Intimate geopolitics provides a method of critically analyzing and understanding how national and international political actions have a significant impact on the everyday lives of individuals and communities. Similarly, the seemingly banal and everyday actions of people can influence and shape political attitudes, policy, and actions at national and even international scales. Therefore, geographic scale remains continually significant for the study of gender and geopolitics. Many contemporary feminist geopolitical analyses include multiple scales of inquiry from the macro (international, regional, and national) to the micro (communities, homes, and bodies). Feminist geographers' examinations of politics at multiple scales elucidate the ways in which daily lives influence, and are influenced by, politics in different places.

Lastly, **Chapter 7** explores how an array of power relations shapes our relationships with the environment within the field of feminist political ecology. This chapter offers students detailed examples of how multiple forms of gendered power shape access to and control over land and other natural resources. These multiple forms of power are also influenced by emotions. Feminist political ecologists are an eclectic group of scholars who align along aspects of social justice and the access to natural resources, such as land and water, and whose concern for human and non-human nature moves from the scale of the body to the globe, privileging the everyday and quotidian struggles for equitable forms of environmental change. Feminist political ecology is an effective tool for international development critique. This standpoint acknowledges that development challenges require attention to multiplicity, recognizing, for instance, that people are not poor or disenfranchised based on one form of power, but multiple factors—both privileges and disadvantages—that shape people's opportunities and well-being in the context of international development practice.

Chapter 8: Conclusion and reflections

Reflections: building solidarity

We began our book with the controversy over SlutWalk, and the need for feminist geography to take seriously the ways in which our lives are shaped by intersectional power—not simply in terms of identity, but as power structures imbued with racial, patriarchal, and heteronormative meanings (among many) that shape space and our human and non-human relationships. The discussion of SlutWalk is a key example of how geographic scholarship benefits from intersectional understandings of power. At the same time, it is also instructive in how feminist solidarity cannot be taken for granted, but requires 'mutual dialogue' and not simply one (dominant) voice (hooks 2013, 57).

Through an intersectional lens, we are hopeful about increased solidarity among feminist geographers and within the discipline more broadly. This kind of solidarity demands that feminism contests racism and homophobia with the same vigor and disdain that it does patriarchy. Such a key message was communicated by feminist scholar Barbara Smith, in her closing speech at the National Women's Studies Association in May 1979. While so long ago, her message begs repeating, as she states:

> The reason racism is a feminist issue is easily explained by the inherent definition of feminism. Feminism is political theory and practice to free all women: women of color, working class women, poor women, physically challenged women, lesbians, old women, as well as white economically privileged heterosexual women. Anything less than this is not feminism, but merely self-aggrandizement.
>
> (Smith, cited in Moraga and Anzaldua 2015, 57)

In this vein, we see growing solidarity in the ways in which our subfields of indigenous, Black, and LGBTQ geographies have overlapping and extended priorities of inclusion that align with feminist geographies in numerous ways, such as (but not limited to) the rejection of positivist myths of impartiality in knowledge production and the importance of historicizing our work to debunk claims of post-coloniality, post-patriarchy, and post-raciality as temporal truths. This solidarity also disrupts dominant notions of our epistemological and ontological conventions that have long governed ways of seeing human and non-human relations. Furthermore, notions of solidarity emerge in the work on race and masculinity, as well as race and gender (attending largely to women and femininity) and the important ways in which feminist geographers are increasingly attentive to gendered racialization, racialized genders, and racialized sexualities as they operate through various environments.

This advance in our disciplinary knowledge production did not happen by accident, but rather through struggle. Such struggles to be heard among these growing subfields are built on a commitment to making visible the plural ways of knowing in our collective fields of study. This most recent advance builds upon the work of feminist geographers, but also is made possible by those geographers engaged with fields and scholarship beyond geography, particularly in the fields of Black studies, African and African American studies, Native studies, indigenous studies, women's and gender studies, and sex and sexuality studies, among others. These fields offer significant sophistication in how we think as geographers.

At the time we write, we are in the early months of a US administration where sexist and racist rhetoric is common, and which uses the law to wield exclusionary policy with negative impacts on mobilities and immigration. These events come at a time when right-wing governments and groups are on the rise both inside and outside the US. For us, this means that the growing scholarly solidarity among feminist, Black, indigenous, and LGBTQ geographic subfields is so very important in a discipline that belongs to no one and everyone simultaneously. Such solidarity and alliances must aim to keep open the diverse, plural, and significant contributions from geographers, a task that will require ever more tenacious insistence in the coming years.

To close, we are reminded of the words of feminist geographer Cindi Katz, which, while written in 1992, could not be more fitting for the future of feminist geography:

> At this juncture we must recognize that in the [face] of diverse and articulated structures of dominance, exploitation, and oppression there are multiple and connected positionings from which to confront and potentially transform these structures. Part of our project as critical intellectuals should be to interrogate these diverse positionings to find common ground for change.
>
> (504–5)

Conclusion and reflections

Recommended reading

hooks, bell. 2013. *Writing Beyond Race: Living Theory and Practice*. New York: Routledge.

Katz, Cindi. 1992. 'All the World Is Staged: Intellectuals and the Projects of Ethnography.' *Environment and Planning D: Society and Space* 10 (5): 495–510.

Moraga, Cherrie, and Gloria Anzaldua. 2015. *This Bridge Called My Back: Writings by Radical Women of Color*. 4th ed. Albany, NY: State University of New York (SUNY) Press.

Smith, Barbara. 1980. 'Racism and Women's Studies.' *Frontiers* 5 (1): 48–9.

Bibliography

Abraham, Janaki. 2010. 'Veiling and the Production of Gender and Space in a Town in North India.' *Indian Journal of Gender Studies* 17 (2): 191–222. doi:10.1177/097152151001700201.

Abu-Lughod, Lila. 1991. 'Writing against Culture.' In *Recapturing Anthropology: Working in the Present*, edited by Richard G. Fox, 137–62. Santa Fe, NM: School of American Research Press.

———. 2013. *Do Muslim Women Need Saving?* Cambridge, MA: Harvard University Press.

Agarwal, Bina. 1994. *A Field of One's Own: Gender and Land Rights in South Asia*. Cambridge, UK: Cambridge University Press.

———. 2003. 'Gender and Land Rights Revisited: Exploring New Prospects via the State, Family and Market.' *Journal of Agrarian Change* 3 (1–2): 184–224.

Ahmed, Sara. 2004. *The Cultural Politics of Emotion*. Edinburgh: Edinburgh University Press.

———. 2006. *Queer Phenomenology: Orientations, Objects, Others*. Durham, NC: Duke University Press.

———. 2010. *The Promise of Happiness*. Durham, NC: Duke University Press.

Alatout, Samer. 2009. 'Bringing Abundance into Environmental Politics: Constructing a Zionist Network of Water Abundance, Immigration, and Colonization.' *Social Studies of Science* 39 (3): 363–94.

Alexander, Michelle. 2012. *The New Jim Crow: Mass Incarceration in the Age of Colorblindness*. New York: The New Press.

Alhayek, Katty. 2014. 'Double Marginalization: The Invisibility of Syrian Refugee Women's Perspectives in Mainstream Online Activism and

Global Media.' *Feminist Media Studies* 14 (4): 696–700. doi:10.1080/14680777.2014.935205.

Alonso-Fradejas, Alberto. 2012. 'Land Control-Grabbing in Guatemala: The Political Economy of Contemporary Agrarian Change.' *Canadian Journal of Development Studies* 33 (4): 509–28.

Alta Gracia Apparel. 2016. 'Alta Gracia Fair Trade: A Living Wage.' www.altagraciaapparel.com (accessed May 23, 2017).

Anthias, Floya, and Nira Yuval-Davis, eds. 1989. *Women–Nation–State*. New York: Palgrave Macmillan.

Arthur, Linda B. 1999. 'Dress and the Social Construction of Gender in Two Sororities.' *Textiles Research Journal* 17 (2): 84–93.

Asher, Kiran. 2009. *Black and Green: Afro-Colombians, Development, and Nature in the Pacific Lowlands*. Durham, NC: Duke University Press.

Bagheri, Nazgol. 2014. 'Mapping Women in Tehran's Public Spaces: A Geo-Visualization Perspective.' *Gender, Place & Culture* 21 (10): 1285–301. doi:10.1080/0966369X.2013.817972.

Baviskar, Amita. 2003. 'Between Violence and Desire: Space, Power, and Identity in the Making of Metropolitan Delhi.' *International Social Science Journal* 55 (175): 89–98. doi:10.1111/1468-2451.5501009.

Beckingham, David. 2012. 'Gender, Space, and Drunkenness: Liverpool's Licensed Premises, 1860–1914.' *Annals of the Association of American Geographers* 102 (3): 647–66. doi:10.1080/00045608.2011.652850.

Bee, Beth A., Jennifer Rice, and Amy Trauger. 2015. 'A Feminist Approach to Climate Change Governance: Everyday and Intimate Politics.' *Geography Compass* 9 (6): 339–50.

Bell, David, and Gill Valentine. 1995. *Mapping Desire: Geographies of Sexualities*. New York: Routledge.

Belser, Patrick. 2005. 'Forced Labour and Human Trafficking: Estimating the Profits.' Geneva, Switzerland: International Labour Organization. http://digitalcommons.ilr.cornell.edu/cgi/viewcontent.cgi?article=1016&context=forcedlabor (accessed July 30, 2017).

Beneria, Lourdes, Gunseli Berik, and Maria S. Floro. 2016. *Gender, Development, and Globalization: Economics As If All People Mattered*. 2nd ed. New York: Routledge.

Berg, Lawrence D., and Robyn Longhurst. 2003. 'Placing Masculinities and Geography.' *Gender, Place & Culture* 10 (4): 351–60. doi:10.1080/0966369032000153322.

Bibliography

Bhavnani, Kum, John Foran, and Priya A. Kurian. 2003. *An Introduction to Women, Culture and Development in Feminist Futures: Re-Imagining Women, Culture and Development*. London: Zed Books.

Black, Michele C., Kathleen C. Basile, Sharon G. Smith, Matthew J. Brieding, Melissa T. Merrick, Mikel L. Walters, Mark R. Stevens, and Jieru Chen. 2011. *The National Intimate Partner and Sexual Violence Survey: 2010 Summary Report*. Atlanta, GA: National Center for Injury Prevention/Centers for Disease Control and Prevention. www.cdc.gov/violenceprevention/pdf/nisvs_report2010-a.pdf (accessed February 27, 2017).

Blunt, Alison, and Robyn Dowling. 2006. *Home*. New York: Routledge.

Blunt, Alison, and Gillian Rose. 1994. *Writing Women and Space: Colonial and Postcolonial Geographies*. New York: Guilford Press.

Bondi, Liz, and Nina Laurie. 2005. 'Introduction.' In *Working the Spaces of Neoliberalism*, edited by Nina Laurie and Liz Bondi, 1–8. Chichester, UK: John Wiley & Sons. http://onlinelibrary.wiley.com/doi/10.1002/9781444397437.ch/summary (accessed July 26, 2017).

Bondi, Liz, and Damaris Rose. 2003. 'Constructing Gender, Constructing the Urban: A Review of Anglo-American Feminist Urban Geography.' *Gender, Place & Culture* 10 (3): 229–45. doi:10.1080/0966369032000114000.

Bosco, Fernando. 2001. 'Place, Space, Networks and the Sustainability of Collective Action.' *Global Networks* 1 (4): 307–29.

Boserup, Ester. 1970. *Women's Role in Economic Development*. New York: St. Martin's Press.

Boyd, Jade. 2010. 'Producing Vancouver's (Hetero)normative Nightscape.' *Gender, Place & Culture* 17 (2): 169–89. doi:10.1080/09663691003600298.

Brickell, Katherine. 2012. '"Mapping" and "Doing" Critical Geographies of Home.' *Progress in Human Geography* 36 (2): 225–44. doi:10.1177/0309132511418708.

Brison, Susan. 2011. 'An Open Letter from Black Women to SlutWalk Organizers.' *Huffington Post*. www.huffingtonpost.com/susan-brison/slutwalk-black-women_b_980215.html (accessed February 27, 2017).

Brown, Michael P. 2000. *Closet Space: Geographies of Metaphor from the Body to the Globe*. London and New York: Routledge.

Browne, Kath. 2004. 'Genderism and the Bathroom Problem: (Re)materialising Sexed Sites, (Re)creating Sexed Bodies.' *Gender, Place & Culture* 11 (3): 331–46. doi:10.1080/0966369042000258668.

———. 2007. 'A Party with Politics? (Re)making LGBTQ Pride Spaces in Dublin and Brighton.' *Social & Cultural Geography* 8 (1): 63–87. doi:10.1080/14649360701251817.

Browne, Kath, and Leela Bakshi. 2011. 'We Are Here to Party? Lesbian, Gay, Bisexual and Trans Leisurescapes Beyond Commercial Gay Scenes.' *Leisure Studies* 30 (2): 179–96. doi:10.1080/02614367.2010.506651.

Brownlow, Alec. 2005. 'A Geography of Men's Fear.' *Geoforum* 36 (5): 581–92. doi:10.1016/j.geoforum.2004.11.005.

Buechler, Stephanie. 2015. 'Climate-Water Challenges and Gendered Adaption Strategies in Rayon, a Riparian Community in Sonora, Mexico.' In *A Political Ecology of Women, Water and Global Environmental Change*, edited by Stephanie Buechler and Anne-Marie Hanson, 99–117. New York: Routledge.

Buechler, Stephanie, and Anne-Marie Hanson. 2015. *A Political Ecology of Women, Water and Global Environmental Change*. New York: Routledge.

Butler, Judith. 1990. *Gender Trouble: Feminism and the Subversion of Identity*. New York: Routledge.

———. 1993. *Bodies that Matter: On the Discursive Limits of Sex*. New York: Routledge.

Cahill, Caitlin. 2006. '"At Risk"? The Fed Up Honeys Re-Present the Gentrification of the Lower East Side.' *Women's Studies Quarterly* 34 (1/2): 334–63.

Carney, Judith. 1996. 'Converting the Wetlands, Engendering the Environment.' In *Liberation Ecologies: Environment, Development, Social Movements*, edited by Richard Peet and Michael Watts, 165–86. New York: Routledge.

Carpenter, R. Charli. 2006. 'Recognizing Gender-Based Violence against Civilian Men and Boys in Conflict Situations.' *Security Dialogue* 37 (1): 83–103. doi:10.1177/0967010606064139.

Caudwell, Jayne, and Kath Browne. 2013. *Sexualities, Spaces and Leisure Studies*. New York: Routledge.

Ceci, Stephen J., Donna K. Ginther, Shulamit Kahn, and Wendy M. Williams. 2014. 'Women in Academic Science: A Changing Landscape.' *Psychological Science in the Public Interest* 15 (3): 75–141.

Chant, Sylvia. 1998. 'Households, Gender and Rural–Urban Migration: Reflections on Linkages and Considerations for Policy.' *Environment and Urbanization* 10 (10): 5–21.

Bibliography

———. 2012. 'The Disappearing of "Smart Economics"? The World Development Report 2012 on Gender Equality: Some Concerns about the Preparatory Process and the Prospects for Paradigm Change.' *Global Social Policy* 12 (2): 1–21.

———. 2013. 'Cities through a "Gender Lens": A Golden "Urban Age" for Women in the Global South?' *Environment and Urbanization* 25 (1): 9–29. doi:10.1177/0956247813477809.

Chant, Sylvia, and Cathy McIlwaine. 2015. *Cities, Slums and Gender in the Global South: Towards a Feminised Urban Future*. New York: Routledge.

Chant, Sylvia, and Caroline Sweetman. 2012. 'Fixing Women or Fixing the World? "Smart Economics," Efficiency Approaches, and Gender Equality in Development.' *Gender and Development* 20 (3): 517–29.

Children's Defense Fund. 2014. 'The State of America's Children.' Washington, DC. www.childrensdefense.org/library/state-of-americas-children/ (accessed February 27, 2017).

Chin, Elizabeth. 2001. *Purchasing Power: Black Kids and American Consumer Culture*. Minneapolis, MN: University of Minnesota Press.

Chouinard, Vera. 1999. 'Body Politics: Disabled Women's Activism in Canada and Beyond.' In *Mind and Body Spaces: Geographies of Illness, Impairment and Disability*, edited by Ruth Butler and Hester Parr, 269–94. New York: Routledge.

Coates, Ta-Nehisi. 2015. *Between the World and Me*. New York: Spiegel & Grau.

Collins, Patricia Hill. 2000. *Black Feminist Thought: Knowledge, Consciousness, and the Politics of Empowerment*. New York: Routledge.

Colls, Rachel. 2004. '"Looking Alright, Feeling Alright": Emotions, Sizing and the Geographies of Women's Experiences of Clothing Consumption.' *Social & Cultural Geography* 5 (4): 583–96.

———. 2006. 'Outsize/Outside: Bodily Bignesses and the Emotional Experiences of British Women Shopping for Clothes.' *Gender, Place & Culture* 13 (5): 529–45.

Connell, R. W. 2005. *Masculinities*. 2nd ed. Berkeley, CA: University of California Press.

Cookson, Tara Patricia. 2016. 'Working for Inclusion? Conditional Cash Transfers, Rural Women, and the Reproduction of Inequality.' *Antipode* 48 (5): 1187–205.

Cope, Meghan. 2004. 'Placing Gendered Political Acts.' In *Mapping Women, Making Politics: Feminist Perspectives on Political Geography*, edited by Lynn

A. Staeheli, Eleonore Kofman, and Linda Peake, 71–86. New York: Routledge.

Cornwall, Andrea, and Jenny Edwards, eds. 2014. *Feminisms, Empowerment and Development: Changing Women's Lives*. London: Zed Books.

Cox, Rosie. 2013. 'House/Work: Home as a Space of Work and Consumption.' *Geography Compass* 7 (12): 821–31. doi:10.1111/gec3.12089.

———. 2015. 'Materials, Skills and Gender Identities: Men, Women and Home Improvement Practices in New Zealand.' *Gender, Place & Culture* 23 (4): 572–88. doi:10.1080/0966369X.2015.1034248.

Crenshaw, Kimberle. 1991. 'Mapping the Margins: Intersectionality, Identity Politics, and Violence against Women of Color.' *Stanford Law Review* 43 (6): 1241–99.

Cuomo, Dana. 2013. 'Security and Fear: The Geopolitics of Intimate Partner Violence Policing.' *Geopolitics* 18 (4): 856–74.

Day, Kristen. 2006. 'Being Feared: Masculinity and Race in Public Space.' *Environment and Planning A* 38 (3): 569–86. doi:10.1068/a37221.

Day, Kristen, Cheryl Stump, and Daisy Carreon. 2003. 'Confrontation and Loss of Control: Masculinity and Men's Fear in Public Space.' *Journal of Environmental Psychology* 23 (3): 311–22.

de Janvry, Alain, and Elisabeth Sadoulet. 2006. 'Making Conditional Cash Transfer Programs More Efficient: Designing for Maximum Effect of the Conditionality.' *The World Bank Economic Review* 20 (1): 1–29.

de Koning, Anouk. 2009. 'Gender, Public Space and Social Segregation in Cairo: Of Taxi Drivers, Prostitutes and Professional Women.' *Antipode* 41 (3): 533–56. doi:10.1111/j.1467-8330.2009.00686.x.

Delaney, David, and Helga Leitner. 1997. 'The Political Construction of Scale.' *Political Geography* 16 (2): 93–97. doi:10.1016/S0962-6298(96)00045-5.

Desmond, Matthew. 2012. 'Eviction and the Reproduction of Urban Poverty.' *American Journal of Sociology* 118 (1): 88–133. doi:10.1086/666082.

Detraz, Nicole. 2012. *International Security and Gender*. Malden, MA: Polity Press.

Dittmer, Jason. 2005. 'Captain America's Empire: Reflections on Identity, Popular Culture, and Post-9/11 Geopolitics.' *Annals of the Association of American Geographers* 95 (3): 626–43. doi:10.1111/j.1467-8306.2005.00478.x.

Dixon, Deborah P., and Sallie A. Marston. 2011. 'Introduction: Feminist Engagements with Geopolitics.' *Gender, Place & Culture* 18 (4): 445–53. doi:10.1080/0966369X.2011.583401.

Doan, Petra. 2009. 'Safety and Urban Environments: Transgendered Experiences of the City.' *Women & Environments International Magazine* 80/81 (Fall/Winter): 22–5.

———. 2010. 'The Tyranny of Gendered Space: Reflections from beyond the Gender Dichotomy.' *Gender, Place & Culture* 17 (5): 635–54. doi:10.1080/0966369X.2010.503121.

Doan, Petra, and Harrison Higgins. 2009. 'Cognitive Dimensions of Way-Finding: The Implications of Habitus, Safety, and Gender Dissonance among Gay and Lesbian Populations.' *Environment and Planning A* 41 (7): 1745–62.

Domosh, Mona, and Joni Seager. 2001. *Putting Women in Place: Feminist Geographers Make Sense of the World.* New York: Guilford Press.

Doshi, Sapana. 2013. 'The Politics of the Evicted: Redevelopment, Subjectivity, and Difference in Mumbai's Slum Frontier.' *Antipode* 45 (4): 844–65. doi:10.1111/j.1467-8330.2012.01023.x.

———. 2017. 'Embodied Urban Political Ecology: Five Propositions.' *Area* 49 (1): 125–8. doi:10.1111/area.12293.

Dowler, Lorraine. 1998. '"And They Think I'm Just a Nice Old Lady": Women and War in Belfast, Northern Ireland.' *Gender, Place & Culture* 5 (2): 159–76.

Dowler, Lorraine, and Joanne Sharp. 2001. 'A Feminist Geopolitics?' *Space and Polity* 5 (3): 165–76.

Durso, Laura E., and Gary J. Gates. 2012. 'Serving Our Youth: Findings from a National Survey of Service Providers Working with Lesbian, Gay, Bisexual and Transgender Youth Who Are Homeless or at Risk of Becoming Homeless.' escholarship.org/uc/item/80x75033.pdf (accessed February 27, 2017).

Economist. 2011 (May 5). 'When Others Are Grabbing Their Land.' www.economist.com/node/18648855 (accessed February 27, 2017).

Elmhirst, Rebecca. 2011. 'Introducing New Feminist Political Ecologies.' *Geoforum* 42 (2): 129–32. doi:10.1016/j.geoforum.2011.01.006.

England, Kim, and Kate Boyer. 2009. 'Women's Work: The Feminization and Shifting Meanings of Clerical Work.' *Journal of Social History* 43 (2): 307–40.

England, Kim, and Caitlin Henry. 2013. 'Care Work, Migration and Citizenship: International Nurses in the UK.' *Social & Cultural Geography* 14 (5): 558–74. doi:10.1080/14649365.2013.786789.

England, Marcia. 2006. 'Breached Bodies and Home Invasions: Horrific Representations of the Feminized Body and Home.' *Gender, Place & Culture* 13 (4): 353–63. doi:10.1080/09663690600808452.

England, Marcia R., and Stephanie Simon. 2010. 'Scary Cities: Urban Geographies of Fear, Difference and Belonging.' *Social & Cultural Geography* 11 (3): 201–7. doi:10.1080/14649361003650722.

Enloe, Cynthia. 1988. *Does Khaki Become You? The Militarization of Women's Lives*. San Francisco, CA: HarperCollins.

———. 1993. *The Morning After: Sexual Politics at the End of the Cold War*. Berkeley, CA: University of California Press.

———. 2000. *Maneuvers: The International Politics of Militarizing Women's Lives*. Berkeley, CA: University of California Press.

Faria, Caroline. 2014. 'Styling the Nation: Fear and Desire in the South Sudanese Beauty Trade.' *Transactions of the Institute of British Geographers* 39 (2): 318–30. doi:10.1111/tran.12027.

Farish, Matthew. 2010. *The Contours of America's Cold War*. Minneapolis, MN: University of Minnesota Press.

Feldman, Roberta M., and Susan Stall. 2004. *The Dignity of Resistance: Women Residents' Activism in Chicago Public Housing*. New York: Cambridge University Press.

Fenster, Tovi. 1999. 'Space for Gender: Cultural Roles of the Forbidden and the Permitted.' *Environment and Planning D: Society and Space* 17 (2): 227–46. doi:10.1068/d170227.

———. 2005. 'The Right to the Gendered City: Different Formations of Belonging in Everyday Life.' *Journal of Gender Studies* 14 (3): 217–31. doi:10.1080/09589230500264109.

Fincher, Ruth, and Jane Margaret Jacobs. 1998. *Cities of Difference*. New York: Guilford Press.

Fluri, Jennifer L. 2006. '"Our Website Was Revolutionary": Virtual Spaces of Representation and Resistance.' *ACME: An International E-Journal for Critical Geographies* 5 (1): 89–111.

———. 2009. '"Foreign Passports Only": Geographies of (Post) Conflict Work in Kabul, Afghanistan.' *Annals of the Association of American Geographers* 99 (5): 986–94.

———. 2011. 'Armored Peacocks and Proxy Bodies: Gender Geopolitics in Aid/Development Spaces of Afghanistan.' *Gender, Place & Culture* 18 (4): 519–36.

Fluri, Jennifer L., and Lorraine Dowler. 2003. 'House Bound: Women's Agency in White Separatist Movements.' In *Spaces of Hate: Geographies of*

Discrimination, and Intolerance in the U.S.A, edited by Colin Flint, 69–86. New York: Routledge.

Fluri, Jennifer L., and Amy Trauger. 2011. 'The Corporeal Marker Project (CMP): Teaching about Bodily Difference, Identity and Place through Experience.' *Journal of Geography in Higher Education* 35 (4): 551–63. doi:10.1080/03098265.2011.552105.

Francis, Elizabeth. 2000. *Making a Living: Changing Livelihoods in Rural Africa*. New York: Routledge.

Freeman, Carla. 2000. *High Tech and High Heels in the Global Economy: Women, Work, and Pink-Collar Identities in the Caribbean*. Durham, NC: Duke University Press.

Friedan, Betty. 1963. *The Feminine Mystique*. New York: Norton.

Gibson-Graham, J. K. 2006. *The End of Capitalism (As We Knew It): A Feminist Critique of Political Economy*. Minneapolis, MN: University of Minnesota Press.

Giddings, Carla, and Alice J. Hovorka. 2010. 'Place, Ideological Mobility and Youth Negotiations of Gender Identities in Urban Botswana.' *Gender, Place & Culture* 17 (2): 211–29. doi:10.1080/09663691003600314.

Gilmore, Ruth Wilson. 2007. *Golden Gulag: Prisons, Surplus, Crisis, and Opposition in Globalizing California*. Berkeley, CA: University of California Press.

Gleeson, Brendan J. 1996. 'A Geography for Disabled People?' *Transactions of the Institute of British Geographers* 21 (2): 387–96.

———. 1999. *Geographies of Disability*. New York: Routledge.

Gökarıksel, Banu, and Anna Secor. 2008. 'New Transnational Geographies of Islamism, Capitalism and Subjectivity: The Veiling-Fashion Industry in Turkey.' *Area* 41 (1): 6–18. doi:10.1111/j.1475-4762.2008.00849.x.

Gorman-Murray, Andrew. 2008a. 'Masculinity and the Home: A Critical Review and Conceptual Framework.' *Australian Geographer* 39 (3): 367–79. doi:10.1080/00049180802270556.

———. 2008b. 'Queering the Family Home: Narratives from Gay, Lesbian and Bisexual Youth Coming Out in Supportive Family Homes in Australia.' *Gender, Place & Culture* 15 (1): 31–44. doi:10.1080/09663690701817501.

Greed, Clara. 2015. 'Taking Women's Bodily Functions into Account in Urban Planning and Sustainability.' In *21st International Sustainability Development Research Society (ISDRS) Conference, The Tipping Point:*

Vulnerability and Adaptive Capacity, Geelong, Australia, July 10-12. Available from: eprints.uwe.ac.uk/25927.

Gregory, Steven. 2014. *The Devil behind the Mirror: Globalization and Politics in the Dominican Republic.* Berkeley, CA: University of California Press.

Gregson, Nicky, and Gillian Rose. 2000. 'Taking Butler Elsewhere: Performativities, Spatialities and Subjectivities.' *Environment and Planning D: Society and Space* 18 (4): 433–52.

Gregson, Nicky, Louise Crewe, and Kate Brooks. 2002. 'Shopping, Space, and Practice.' *Environment and Planning D: Society and Space* 20 (5): 597–617. doi:10.1068/d270t.

Gururani, Shubhra. 2002. 'Forests of Pleasure and Pain: Gendered Practices of Labor and Livelihood in the Forests of the Kumaon Himalayas, India.' *Gender, Place & Culture* 9 (3): 229–43. doi:10.1080/0966369022000003842.

Hale, Angela, and Jane Wills. 2007. 'Women Working Worldwide: Transnational Networks, Corporate Social Responsibility and Action Research.' *Global Networks* 7 (4): 453–76.

Hansen, Nancy, and Chris Philo. 2007. 'The Normality of Doing Things Differently: Bodies, Spaces and Disability Geography.' *Tijdschrift Voor Economische En Sociale Geografie* 98 (4): 493–506. doi:10.1111/j.1467-9663.2007.00417.x.

Hanson, Susan, and Geraldine Pratt. 1991. 'Job Search and the Occupational Segregation of Women.' *Annals of the Association of American Geographers* 81 (2): 229–53.

———. 1995. *Gender, Work and Space.* New York: Routledge.

Hawkins, Roberta, and Diana Ojeda. 2011. 'Gender and Environment: Critical Tradition and New Challenges.' *Environment and Planning D: Society and Space* 29 (2): 237–53.

Hayden, Dolores. 1980. 'What Would a Non-Sexist City Be Like? Speculations on Housing, Urban Design, and Human Work.' *Signs: Journal of Women in Culture and Society* 5 (3): S170–87.

Hegewisch, Ariane, and Heidi Hartmann. 2015. 'The Gender Wage Gap: 2014.' Washington, DC: Institute for Women's Policy Research. www.iwpr.org/publications/pubs/the-gender-wage-gap-2014 (accessed February 27, 2017).

Henderson, Victoria L. 2009. 'Citizenship in the Line of Fire: Protective Accompaniment, Proxy Citizenship, and Pathways for Transnational Solidarity in Guatemala.' *Annals of the Association of American Geographers* 99 (5): 969–76.

Hill, Catherine, Christianne Corbett, and Andresse St. Rose. 2010. 'Why So Few? Women in Science, Technology, Engineering, and Mathematics.' *The American Association of University Women.* www.aauw.org (accessed February 27, 2017).

Hofman, Nila. 2013. 'Working the Yucatán in the Neoliberal Landscape: Household Management, Consumption and the Lives of Intellectual Elite Women.' *Gender, Place & Culture* 20 (8): 999–1014. doi:10.1080/0966369X.2012.737769.

Holliday, Ruth, David Bell, Meredith Jones, Kate Hardy, Emily Hunter, Elspeth Probyn, and Jacqueline Sanchez Taylor. 2015. 'Beautiful Face, Beautiful Place: Relational Geographies and Gender in Cosmetic Surgery Tourism Websites.' *Gender, Place & Culture* 22 (1): 90–106. doi:10.1080/0966369X.2013.832655.

hooks, bell. 1990. *Yearning: Race, Gender, and Cultural Politics.* Boston, MA: South End Press.

———. 2000. *Feminist Theory: From Margin to Center.* London: Pluto Press.

———. 2013. *Writing Beyond Race: Living Theory and Practice.* New York: Routledge.

Hovorka, Alice J. 2012. 'Women/Chickens vs. Men/Cattle: Insights on Gender–Species Intersectionality.' *Geoforum* 43 (4): 875–84. doi:10.1016/j.geoforum.2012.02.005.

———. 2015. 'The *Gender, Place & Culture* Jan Monk Distinguished Annual Lecture: Feminism and Animals: Exploring Interspecies Relations through Intersectionality, Performativity and Standpoint.' *Gender, Place & Culture* 22 (1): 1–19.

Huang, Shirlena, and Brenda S. A. Yeoh. 2008. 'Heterosexualities and the Global(ising) City in Asia: Introduction.' *Asian Studies Review* 32 (1): 1–6. doi:10.1080/10357820701871187.

Hubbard, Phil. 2004. 'Revenge and Injustice in the Neoliberal City: Uncovering Masculinist Agendas.' *Antipode* 36 (4): 665–86. doi:10.1111/j.1467-8330.2004.00442.x.

———. 2013. 'Carnage! Coming to a Town near You? Nightlife, Uncivilised Behaviour and the Carnivalesque Body.' *Leisure Studies* 32 (3): 265–82. doi:10.1080/02614367.2011.633616.

Hyndman, Jennifer. 2012. 'The Geopolitics of Migration and Mobility.' *Geopolitics* 17 (2): 243–55. doi:10.1080/14650045.2011.569321.

IM-Defensoras. 2015. 'Iniciativa Mesoamericana de Mujeres Defensoras de Derechos Humanos.' *Iniciativa Mesoamericana de Mujeres Defensoras de Derechos Humanos.* im-defensoras.org/es/ (accessed February 27, 2017).

Imrie, Rob. 2004. 'Disability, Embodiment and the Meaning of the Home.' *Housing Studies* 19 (5): 745–64.

Jackson, Peter. 1989. *Maps of Meaning: An Introduction to Cultural Geography.* New York: Routledge.

Jacobs, Jane M., and Catherine Nash. 2003. 'Too Little, Too Much: Cultural Feminist Geographies.' *Gender, Place & Culture* 10 (3): 265–79. doi:10.1080/0966369032000114037.

Jacobs, Jessica. 2009. 'Have Sex Will Travel: Romantic "Sex Tourism" and Women Negotiating Modernity in the Sinai.' *Gender, Place & Culture* 16 (1): 43–61. doi:10.1080/09663690802574787.

———. 2010. *Sex, Tourism and the Postcolonial Encounter: Landscapes of Longing in Egypt.* New York: Routledge.

Jarosz, Lucy. 1991. 'Women as Rice Sharecroppers in Madagascar.' *Society & Natural Resources* 4 (1): 53–63.

Jin, Xiuming, and Risa Whitson. 2014. 'Young Women and Public Leisure Spaces in Contemporary Beijing: Recreating (with) Gender, Tradition, and Place.' *Social & Cultural Geography* 15 (4): 449–69. doi:10.1080/14649365.2014.894115.

Johnston, Lynda. 1996. 'Flexing Femininity: Female Body-Builders Refiguring "the Body."' *Gender, Place & Culture* 3 (3): 327–40. doi:10.1080/09663699625595.

———. 2005. *Queering Tourism: Paradoxical Performances at Gay Pride Parades.* New York: Routledge.

Johnston, Lynda, and Robyn Longhurst. 2010. *Space, Place, and Sex: Geographies of Sexualities.* New York: Rowman & Littlefield.

Kabeer, Naila. 1994. *Reversed Realities: Gender Hierarchies in Development Thought.* New York: Verso.

Kabeer, Naila, and Simeen Mahmud. 2004. 'Globalization, Gender and Poverty: Bangladeshi Women Workers in Export and Local Markets.' *Journal of International Development* 16 (1): 93–109.

Kandiyoti, Deniz. 1988. 'Bargaining with Patriarchy.' *Gender and Society* 2 (3): 274–90.

Katz, Cindi. 1992. 'All the World Is Staged: Intellectuals and the Projects of Ethnography.' *Environment and Planning D: Society and Space* 10 (5): 495–510.

———. 1998. 'Excavating the Hidden City of Social Reproduction: A Commentary.' *City & Society* 10 (1): 37–46. doi:10.1525/city.1998.10.1.37.

———. 2001. 'Vagabond Capitalism and the Necessity of Social Reproduction.' *Antipode* 33 (4): 709–28.

———. 2004. *Growing Up Global: Economic Restructuring and Children's Everyday Lives*. Minneapolis, MN: University of Minnesota Press.

Kern, Leslie. 2010. 'Selling the "Scary City": Gendering Freedom, Fear and Condominium Development in the Neoliberal City.' *Social & Cultural Geography* 11 (3): 209–30. doi:10.1080/14649361003637174.

———. 2013. 'All Aboard? Women Working the Spaces of Gentrification in Toronto's Junction.' *Gender, Place & Culture* 20 (4): 510–27. doi:10.1080/0966369X.2012.701201.

Klein, Naomi. 1999. *No Logo: Taking Aim at the Brand Bullies*. Toronto: Knopf.

Kobayashi, Audrey. 2014. 'The Dialectic of Race and the Discipline of Geography.' *Annals of the Association of American Geographers* 104 (6): 1101–15.

Kofman, Eleonore, and Linda Peake. 1990. 'Into the 1990s: A Gendered Agenda for Political Geography.' *Political Geography Quarterly* 9 (4): 313–36.

Koopman, Sara. 2011. 'Alter-Geopolitics: Other Securities Are Happening.' *Geoforum* 42 (3): 274–84. doi:10.1016/j.geoforum.2011.01.007.

Koskela, Hille. 1997. '"Bold Walk and Breakings": Women's Spatial Confidence versus Fear of Violence.' *Gender, Place & Culture* 4 (3): 301–20. doi:10.1080/09663699725369.

Law, Lisa. 1997. 'Dancing on the Bar: Sex, Money and the Uneasy Politics of Third Space.' In *Geographies of Resistance*, edited by Michael Keith and Steven Pile, 107–23. New York: Routledge.

Lawson, Victoria. 2007. *Making Development Geography*. New York: Routledge.

Lefebvre, Henri. 1996. *Writings on Cities*. Translated by Eleonore Kofman and Elizabeth Lebas. Oxford: Blackwell.

Leszczynski, Agnieszka, and Sarah Elwood. 2015. 'Feminist Geographies of New Spatial Media.' *The Canadian Geographer* 59 (1): 12–28.

Lim, Jason, and Alexandra Fanghanel. 2013. '"Hijabs, Hoodies and Hotpants": Negotiating the "Slut" in SlutWalk.' *Geoforum* 48: 207–15.

Longhurst, Robyn. 2008. *Maternities: Gender, Bodies and Space*. New York: Routledge.

Loyd, Jenna, Matt Mitchelson, and Andrew Burridge, eds. 2012. *Beyond Walls and Cages: Prisons, Borders, and Global Crisis*. Athens, GA: University of Georgia Press.

Lyon, Sarah, Josefina Aaranda Bezaury, and Tad Mutersbaugh. 2010. 'Gender Equity in Fairtrade – Organic Coffee Producer Organizations: Cases from Mesoamerica.' *Geoforum* 41 (1): 93–103.

Maalsen, Sophia, and Jessica McLean. 2015. 'Digging Up Unearthed Down-Under: A Hybrid Geography of a Musical Space that Essentialises Gender and Place.' *Gender, Place & Culture* 23 (3): 418–34. doi:10.1080/0966369X.2015.1013443.

Mackenzie, Fiona D. 2010. 'Gender, Land Tenure and Globalisation: Exploring Conceptual Ground.' In *Land Tenure, Gender and Globalisation: Research and Analysis from Africa, Asia and Latin America*, edited by Dzodzi Tsikata and Pamela Golah, 35–50. Ottawa: International Development Research Center.

Madge, Clare, and Henrietta O'Connor. 2005. 'Mothers in the Making? Exploring Liminality in Cyber/Space.' *Transactions of the Institute of British Geographers* 30 (1): 83–97. doi:10.1111/j.1475-5661.2005.00153.x.

Mahtani, Minelle. 2014. 'Toxic Geographies: Absences in Critical Race Thought and Practice in Social and Cultural Geography.' *Social & Cultural Geography* 15 (4): 359–67.

Mainardi, Pat. 2012. 'The Politics of Housework.' In *Women: Images and Realities: A Multicultural Anthology*, edited by Suzanne Kelly, Gowri Parameswaran, and Nancy Schniedewind, 185–7. New York: McGraw-Hill.

Maldonado, Marta Maria, Adela C. Licona, and Sarah Hendricks. 2016. 'Latin@ Immobilities and Altermobilities within the U.S. Deportability Regime.' *Annals of the American Association of Geographers* 106 (2): 321–9.

Marshall, David Jones. 2014. 'Love Stories of the Occupation: Storytelling and the Counter-Geopolitics of Intimacy.' *Area* 46 (4): 349–51.

Marston, Sallie A., John Paul Jones III, and Keith Woodward. 2005. 'Human Geography without Scale.' *Transactions of the Institute of British Geographers* 30 (4): 416–32.

Martin, Lauren. 2011. 'The Geopolitics of Vulnerability: Children's Legal Subjectivity, Immigrant Family Detention and US Immigration Law and Enforcement Policy.' *Gender, Place & Culture* 18 (4): 477–98. doi:10.1080/0966369X.2011.583345.

Martin, Patricia M., and Nohora Carvajal. 2015. 'Feminicide as "Act" and "Process": A Geography of Gendered Violence in Oaxaca.' *Gender, Place & Culture* 23 (7): 1–14. doi:10.1080/0966369X.2015.1073693.

Massey, Doreen. 1994. *Space, Place, and Gender*. Minneapolis, MN: University of Minnesota Press.

———. 1995. *Spatial Divisions of Labor: Social Structures and the Geography of Production*. 2nd ed. New York: Routledge.

———. 2005. *For Space*. London: Sage.

Mayer, Tamar. 2012. 'The Struggle over Boundary and Memory: Nation, Borders, and Gender in Jewish Israel.' *Journal of International Women's Studies* 13 (4): 29–50.

McClintock, Anne. 1993. 'Family Feuds: Gender, Nationalism and the Family.' *Feminist Review* 44: 61–80.

McCormack, Derek. 1999. 'Body Shopping: Reconfiguring Geographies of Fitness.' *Gender, Place & Culture* 6 (2): 155–77. doi:10.1080/09663699925088.

McDowell, Linda. 1991. 'The Baby and the Bath Water: Diversity, Deconstruction and Feminist Theory in Geography.' *Geoforum* 22 (2): 123–33.

———. 1997. *Capital Culture: Gender at Work in the City*. Oxford: Blackwell.

———. 2003. *Redundant Masculinities? Employment Change and White Working Class Youth*. Oxford: Blackwell.

———. 2013. *Working Lives: Gender, Migration and Employment in Britain, 1945–2007*. Oxford: Wiley-Blackwell.

McFarlane, Colin, Renu Desai, and Steve Graham. 2014. 'Informal Urban Sanitation: Everyday Life, Poverty, and Comparison.' *Annals of the Association of American Geographers* 104 (5): 989–1011. doi:10.1080/00045608.2014.923718.

McKittrick, Katherine. 2006. *Demonic Grounds: Black Women and the Cartographies of Struggle*. Minneapolis, MN: University of Minnesota Press.

———. 2011. 'On Plantations, Prisons, and a Black Sense of Place.' *Social & Cultural Geography* 12 (8): 947–63. doi:10.1080/14649365.2011.624280.

McKittrick, Katherine, and Linda Peake. 2005. 'What Difference Does Difference Make to Geography?' In *Questioning Geography: Fundamental Debates*, edited by Noel Castree, Alisdair Rogers, and Douglas Sherman, 39–54. Oxford: Blackwell.

McKittrick, Katherine, and Clyde Woods, eds. 2007. *Black Geographies and the Politics of Place*. Cambridge, MA: South End Press.

McLean, Jessica, and Sophia Maalsen. 2013. 'Destroying the Joint and Dying of Shame? A Geography of Revitalised Feminism in

Social Media and Beyond.' *Geographical Research* 51 (3): 243–56. doi:10.1111/1745-5871.12023.

Meehan, Katie, and Kendra Strauss. 2015. *Precarious World: Contested Geographies of Social Reproduction*. Athens, GA: University of Georgia Press.

Mies, Maria. 1982. *The Lace Makers of Narsapur: Indian Housewives Produce for the World Market*. London: Zed Books.

———. 1986. *Patriarchy and Accumulation on a World Scale: Women in the International Division of Labour*. London: Zed Books.

Mies, Maria, Veronika Bennholdt-Thomsen, and Claudia von Werlhof. 1988. *Women: The Last Colony*. London: Zed Books.

Millar, Sarah. 2011 (March 17). 'Police Officer's Remarks at York Inspire "SlutWalk."' *Toronto Star*. www.thestar.com/news/gta/2011/03/17/police_officers_remarks_at_york_inspire_slutwalk.html (accessed August 3, 2017).

Mitchell, Don. 1995. 'The End of Public Space? People's Park, Definitions of the Public, and Democracy.' *Annals of the American Association of Geographers* 85 (1): 108–33.

Mitchell, Katharyne, Sallie A. Marston, and Cindi Katz, eds. 2004. *Life's Work: Geographies of Social Reproduction*. Malden, MA: Blackwell.

Moghadam, Valentine M. 2005. *Globalizing Women: Transnational Feminist Networks*. Baltimore, MD: Johns Hopkins University Press, 2005.

Mohammad, Robina. 2013. 'Making Gender Ma(r)king Place: Youthful British Pakistani Muslim Women's Narratives of Urban Space.' *Environment and Planning A* 45 (8): 1802–22. doi:10.1068/a45253.

Mohanty, Chandra Talpade. 1986. 'Under Western Eyes: Feminist Scholarship and Colonial Discourses.' *Boundary* 12 (3): 333–58.

———. 1991. *Third World Women and the Politics of Feminism*. Bloomington, IN: Indiana University Press.

———. 2003a. *Feminism without Borders: Decolonizing Theory, Practicing Solidarity*. Durham, NC: Duke University Press.

———. 2003b. '"Under Western Eyes" Revisited: Feminist Solidarity through Anticapitalist Struggles.' *Signs: Journal of Women in Culture and Society* 28 (2): 499–535.

Mollett, Sharlene. 2010. 'Esta Listo (Are You Ready)? Gender, Race and Land Registration in the Rio Plátano Biosphere Reserve.' *Gender, Place & Culture* 17 (3): 357–75.

———. 2011. 'Racial Narratives: Miskito and Colono Land Struggles in the Honduran Mosquitia.' *Cultural Geographies* 18 (1): 43–62.

———. 2015. '"Displaced Futures": Indigeneity, Land Struggle, and Mothering in Honduras.' *Politics, Groups, and Identities* 3 (4): 678–83.

———. 2016. 'The Power to Plunder: Rethinking Land Grabbing in Latin America.' *Antipode* 48 (2): 412–32.

Mollett, Sharlene, and Caroline Faria. 2013. "Messing with Gender in Feminist Political Ecology." *Geoforum* 45: 116–25. doi:10.1016/j.geoforum.2012.10.009.

———. Forthcoming. 'The Spatialities of Intersectional Thinking.' *Gender, Place & Culture*.

Momsen, Janet H., and Janet Townsend. 1987. *Geography of Gender in the Third World*. London: Hutchinson.

Monk, Janice, and Susan Hanson. 1982. 'On Not Excluding Half of the Human in Human Geography.' *The Professional Geographer* 34 (1): 11–23. doi:10.1111/j.0033-0124.1982.00011.x.

Moraga, Cherrie, and Gloria Anzaldua. 2015. *This Bridge Called My Back: Writings by Radical Women of Color*. 4th ed. Albany, NY: State University of New York (SUNY) Press.

Morin, Karen M., Robyn Longhurst, and Lynda Johnston. 2001. '(Troubling) Spaces of Mountains and Men: New Zealand's Mount Cook and Hermitage Lodge.' *Social & Cultural Geography* 2 (2): 117–39. doi:10.1080/14649360122194.

Morris, Carol, and Nick Evans. 2001. '"Cheese Makers Are Always Women": Gendered Representations of Farm Life in the Agricultural Press.' *Gender, Place & Culture* 8 (4): 375–90. doi:10.1080/09663690120111618.

Morrison, Carey-Ann. 2013. 'Homemaking in New Zealand: Thinking through the Mutually Constitutive Relationship between Domestic Material Objects, Heterosexuality and Home.' *Gender, Place & Culture* 20 (4): 413–31. doi:10.1080/0966369X.2012.694358.

Moss, Pamela, and Karen Falconer Al-Hindi. 2008. *Feminisms in Geography: Rethinking Space, Place, and Knowledges*. Lanham, MD.: Rowman & Littlefield.

Mountz, Alison. 2004. 'Embodying the Nation-State: Canada's Response to Human Smuggling.' *Political Geography* 23 (3): 323–45.

———. 2011. 'Where Asylum-Seekers Wait: Feminist Counter-Topographies of Sites between States.' *Gender, Place & Culture* 18 (3): 381–99. doi:10.1080/0966369X.2011.566370.

Muller, Tiffany K. 2007. 'Liberty for All? Contested Spaces of Women's Basketball.' *Gender, Place & Culture* 14 (2): 197–213. doi:10.1080/09663690701213776.

Mullings, Beverly. 1999. 'Globalization, Tourism, and the International Sex Trade.' In *Sun, Sex, and Gold: Tourism and Sex Work in the Caribbean*, edited by Kamala Kempadoo, 55–80. New York: Rowman & Littlefield.

Nagar, Richa. 2002. 'Footloose Researchers, "Traveling" Theories, and the Politics of Transnational Feminist Praxis.' *Gender, Place & Culture* 9 (2): 179–86. doi:10.1080/09663960220139699.

———. 2014. *Muddying the Waters: Coauthoring Feminism across Scholarship and Activism*. Urbana and Springfield, IL: University of Illinois Press.

Najjar, Dina. 2015. 'Women's Contributions to Climate Change Adaption in Egypt's Mubarak Resettlement Scheme through Cactus Cultivation and Adjusted Irrigation.' In *A Political Ecology of Women, Water and Global Environmental Change*, edited by Stephanie Buechler and Anne-Marie Hanson, 141–62. New York: Routledge.

Narayan, Uma. 1997. *Dislocating Cultures: Identities, Traditions, and Third-World Feminism*. New York: Routledge.

Nash, Catherine. 2008. *Of Irish Descent: Origin Stories, Genealogy, & the Politics of Belonging*. New York: Syracuse University Press.

———. 2013. 'Queering Neighbourhoods: Politics and Practice in Toronto.' *ACME: An International E-Journal for Critical Geographies* 12 (2): 193–213.

Nash, Catherine J., and Andrew Gorman-Murray. 2014. 'LGBT Neighbourhoods and "New Mobilities": Towards Understanding Transformations in Sexual and Gendered Urban Landscapes.' *International Journal of Urban and Regional Research* 38 (3): 756–72. doi:10.1111/1468-2427.12104.

National Academy of Sciences, National Academy of Engineering, and Institute of Medicine. Committee on Maximizing the Potential of Women in Academic Science and Engineering. 2007. *Beyond Bias and Barriers: Fulfilling the Potential of Women in Academic Science and Engineering*. Washington, DC: The National Academies Press.

Nelson, Ingrid. 2012. 'A Feminist Political Ecology of Livelihoods and Intervention in the Miombo Woodlands of Zambézia, Mozambique.' Dissertation, Eugene, OR: University of Oregon.

Nelson, Lise. 1999. 'Bodies (and Spaces) Do Matter: The Limits of Performativity.' *Gender, Place & Culture* 6 (4): 331–53. doi:10.1080/09663699924926.

Nelson, Lise, and Joni Seager, eds. 2005. *A Companion to Feminist Geography*. Hoboken, NJ: John Wiley & Sons.

Nightingale, Andrea J. 2011. 'Bounding Difference: Intersectionality and the Material Production of Gender, Caste, Class and Environment in Nepal.' *Geoforum* 42 (2): 153–62. doi:10.1016/j.geoforum.2010.03.004.

Njeru, Jeremia N., Ibipo Johnston-Anumonwo, and Samuel Owuor. 2014. 'Gender Equity and Commercialization of Public Toilet Services in Nairobi, Kenya.' In *Global Perspectives on Gender and Space: Engaging Feminism and Development*, edited by Ibipo Johnston-Anumonwo and Ann M. Oberhauser, 17–34. London: Routledge.

Nolan, Barry, and Garland Waller. 2011. *No Way Out But One*. Documentary.

Norris, Pippa. 1985. 'Women's Legislative Participation in Western Europe.' *West European Politics* 8 (4): 90–101.

Oberhauser, Ann M. 2000. 'Feminism and Economic Geography: Gendering Work and Working Gender.' In *A Companion to Economic Geography*, edited by Eric Sheppard and Trevor J. Barnes, 60–76. Oxford: Blackwell.

———. 2002. 'Relocating Gender and Rural Economic Strategies.' *Environment and Planning A* 34 (7): 1221–37. doi:10.1068/a34224.

———. 2016. '(Re)constructing Rural–Urban Spaces: Gendered Livelihoods, Migration and Natural Resources in South Africa.' *GeoJournal* 81 (3): 489–502. doi:10.1007/s10708-015-9635-5.

Oswin, Natalie. 2008. 'Critical Geographies and the Uses of Sexuality: Deconstructing Queer Space.' *Progress in Human Geography* 32 (1): 89–103.

Oxfam International. 2016 (September 26). 'Murder and Eviction: The Global Land Rush Enters New More Violent Phase.' *Oxfam International*. www.oxfam.org/en/pressroom/pressreleases/2016-09-26/murder-and-eviction-global-land-rush-enters-new-more-violent (accessed February 27, 2017).

Oza, Rupal. 2006. *The Making of Neoliberal India: Nationalism, Gender, and the Paradoxes of Globalization*. New York: Routledge.

Pain, Rachel. 2001. 'Gender, Race, Age and Fear in the City.' *Urban Studies* 38 (5/6): 899–913. doi:10.1080/00420980120046590.

———. 2014. 'Gendered Violence: Rotating Intimacy.' *Area* 46 (4): 351–3.

Pain, Rachel, and Susan Smith. 2008. *Fear: Critical Geopolitics and Everyday Life*. Hampshire, UK: Ashgate.

Pain, Rachel, and Lynn Staeheli. 2014. 'Introduction: Intimacy-Geopolitics and Violence.' *Area* 46 (4): 344–7. doi:10.1111/area.12138.

Panelli, Ruth, Anna Kraack, and Jo Little. 2005. 'Claiming Space and Community: Rural Women's Strategies for Living With, and Beyond, Fear.' *Geoforum* 36 (4): 495–508. doi:10.1016/j.geoforum.2004.08.002.

Papanek, Hanna, and Gail Minault, eds. 1982. *Separate Worlds: Studies of Purdah in South Asia*. Columbia, MO: South Asia Books.

Parameswaran, Radhika, and Kavitha Cardoza. 2009. 'Melanin on the Margins: Advertising and the Cultural Politics of Fair/Light/White Beauty in India.' *Journalism & Communication Monographs* 11 (3): 213–74. doi:10.1177/152263790901100302.

Parks, Virginia. 2016. 'Rosa Parks Redux: Racial Mobility Projects on the Journey to Work.' *Annals of the American Association of Geographers* 106 (2): 292–9.

Patel, Reena. 2010. *Working the Night Shift: Women in India's Call Center Industry*. Stanford, CA: Stanford University Press.

Pavlovskaya, Marianna, and Kevin St. Martin. 2007. 'Feminism and Geographic Information Systems: From a Missing Object to a Mapping Subject.' *Geography Compass* 1/3: 583–606. doi:10.1111/j.1749-8198.2007.0028.x.

Peake, Linda, and Martina Rieker, eds. 2013. *Rethinking Feminist Interventions into the Urban*. New York: Routledge.

Pearson, Ruth. 2007. 'Reassessing Paid Work and Women's Empowerment: Lessons from the Global Economy.' In *Feminisms in Development: Contradictions, Contestations and Challenges*, edited by Andrea Cornwall, Elizabeth Harrison, and Ann Whitehead, 201–13. London: Zed Books.

Peck, Jamie, Nikolas Theodore, and Neil Brenner. 2009. 'Neoliberal Urbanism: Models, Moments, Mutations.' *SAIS Review* 29 (1): 49–66.

Penner, Barbara. 2013. 'Researching Female Public Toilets: Gendered Spaces, Disciplinary Limits.' *Journal of International Women's Studies* 6 (2): 81–98.

Perrons, Diane. 2004. 'Understanding Social and Spatial Divisions in the New Economy: New Media Clusters and the Digital Divide.' *Economic Geography* 80 (1): 45–61.

Perry, Keisha-Khan Y. 2009. '"If We Didn't Have Water": Black Women's Struggle for Urban Land Rights in Brazil.' *Environmental Justice* 2 (1): 9–14.

———. 2013. *Black Women against the Land Grab: The Fight for Racial Justice in Brazil*. Minneapolis, MN: University of Minnesota Press.

Phadke, Shilpa. 2005. '"You Can Be Lonely in a Crowd": The Production of Safety in Mumbai.' *Indian Journal of Gender Studies* 12 (1): 41–62. doi:10.1177/097152150401200102.

Phadke, Shilpa, Sameera Khan, and Shilpa Ranade. 2011. *Why Loiter? Women and Risk on Mumbai Streets*. New Delhi: Penguin Books India.

Pilkey, Brent. 2014. 'Queering Heteronormativity at Home: Older Gay Londoners and the Negotiation of Domestic Materiality.' *Gender, Place & Culture* 21 (9): 1142–57. doi:10.1080/0966369X.2013.832659.

Pollard, Jane. 2013. 'Gendering Capital: Financial Crisis, Financialization and (an Agenda for) Economic Geography.' *Progress in Human Geography* 37 (3): 403–23. doi:10.1177/0309132512462270.

Pratt, Geraldine. 1993. 'Reflections on Poststructuralism and Feminist Empirics, Theory and Practice.' *Antipode* 25 (1): 51–63. doi:10.1111/j.1467-8330.1993.tb00216.x.

———. 2004. *Working Feminism*. Philadelphia, PA: Temple University Press.

———. 2005. 'From Migrant to Immigrant: Domestic Workers Settle in Vancouver, Canada.' In *A Companion to Feminist Geography*, edited by Lise Nelson and Joni Seager, 123–37. Hoboken, NJ: John Wiley & Sons.

———. 2012. *Families Apart: Migrant Mothers and the Conflicts of Labor and Love*. Minneapolis, MN: University of Minnesota Press.

Pratt, Geraldine, and Susan Hanson. 1988. 'Gender, Class and Space.' *Environment and Planning D: Society and Space* 6 (1): 15–35.

Pratt, Geraldine, and Victoria Rosner. 2012. *The Global and the Intimate: Feminism in Our Time*. New York: Columbia University Press.

Pratt, Geraldine, and Brenda Yeoh. 2003. 'Transnational (Counter) Topographies.' *Gender, Place & Culture* 10 (2): 159–66.

Puar, Jasbir K. 2007. *Terrorist Assemblages: Homonationalism in Queer Times*. Durham, NC: Duke University Press.

———. 2013. 'Rethinking Homonationalism.' *International Journal of Middle East Studies* 45 (2): 336–9. doi:http://dx.doi.org/10.1017/S002074381300007X.

Pulido, Laura. 2000. 'Rethinking Environmental Racism: White Privilege and Urban Development in Southern California.' *Annals of the American Association of Geographers* 90 (1): 12–40.

———. 2006. *Black, Brown, Yellow, and Left: Radical Activism in Los Angeles*. Berkeley, CA: University of California Press.

———. 2016. 'Geographies of Race and Ethnicity II: Environmental Racism, Racial Capitalism and State-Sanctioned Violence.' *Progress in Human Geography* 41 (4): 1–10. doi:10.1177/0309132516646495.

Purcell, Mark. 2002. 'Excavating Lefebvre: The Right to the City and Its Urban Politics of the Inhabitant.' *GeoJournal* 58 (2): 99–108.

Puwar, Nirmal. 2004. *Space Invaders: Race, Gender and Bodies out of Place.* New York: Berg.

Qian, Junxi, Lei Wei, and Hong Zhu. 2012. 'Consuming the Tourist Gaze: Imaginative Geographies and the Reproduction of Sexuality in Lugu Lake.' *Geografiska Annaler: Series B, Human Geography* 94 (2): 107–24. doi:10.1111/j.1468-0467.2012.00399.x.

Rabbiosi, Chiara. 2014. 'The Condition of "Cosmo-Housewives": Leisure Shopping, the Mainstream and Its Ambiguities.' *Gender, Place & Culture* 21 (2): 211–27.

Radcliffe, Sarah. 2015. *Dilemmas of Difference: Indigenous Women and the Limits of Postcolonial Development Policy.* Durham, NC: Duke University Press.

Radcliffe, Sarah, and Andrea Pequeño. 2010. 'Ethnicity, Development and Gender: Tsáchila Indigenous Women in Ecuador.' *Development and Change* 41 (6): 983–1016.

Radel, Claudia, Birgit Schmook, Nora Haenn, and Lisa Green. 2016. 'The Gender Dynamics of Conditional Cash Transfers and Smallholder Farming in Calakmul, Mexico.' *Women's Studies International Forum.* doi:/10.1016/j.wsif.2016.06.004.

Rai, Shirin M. 2013. *Gender and the Political Economy of Development: From Nationalism to Globalization.* Malden, MA: Blackwell.

Rajan, Ramkishen S., and Sadhana Srivastava. 2007. 'Global Outsourcing of Services: Issues and Implications.' *Harvard Asia Pacific Review* 9 (1): 39–40.

Rocheleau, Dianne E., Barbara P. Thomas-Slayter, and Esther Wangari, eds. 1996. *Feminist Political Ecology: Global Issues and Local Experiences.* New York: Routledge.

Rose, Gillian. 1993. *Feminism and Geography: The Limits of Geographical Knowledge.* Minneapolis, MN: University of Minnesota Press.

———. 2016. 'Rethinking the Geographies of Cultural "Objects" through Digital Technologies: Interface, Network and Friction.' *Progress in Human Geography* 40 (3): 334–51. doi:10.1177/0309132515580493.

Rosewarne, Lauren. 2005. 'The Men's Gallery – Outdoor Advertising and Public Space: Gender, Fear, and Feminism.' *Women's Studies International Forum* 28 (1): 67–78. doi:10.1016/j.wsif.2005.02.005.

Sabhlok, Anu. 2010. 'National Identity in Relief.' *Geoforum* 41 (5): 743–51.

Samarasinghe, Vidyamali. 2012. *Female Sex Trafficking in Asia: The Resilience of Patriarchy in a Changing World.* New York: Routledge.

Sasser, Jade S. 2014. 'The Wave of the Future? Youth Advocacy at the Nexus of Population and Climate Change.' *The Geographical Journal* 180 (2): 102–10.

Scott, James C. 1985. *Weapons of the Weak: Everyday Forms of Peasant Resistance*. New Haven, CT: Yale University Press.

Seager, Joni. 1993. *Earth Follies: Coming to Feminist Terms with the Global Environmental Crisis*. New York: Routledge.

———. 2003. 'Rachel Carson Died of Breast Cancer: The Coming of Age of Feminist Environmentalism.' *Signs: Journal of Women in Culture and Society* 28 (3): 945–72.

Secor, Anna. 2004. '"There Is an Istanbul that Belongs to Me": Citizenship, Space, and Identity in the City.' *Annals of the Association of American Geographers* 94 (2): 352–68.

Sen, Gita, and Caren Grown. 1985. *Development Alternatives with Women for a New Era*. London: Earthscan.

Shabazz, Rashad. 2012. 'Mapping Black Bodies for Disease: Prisons, Migration, and the Politics of HIV/AIDS.' In *Beyond Walls and Cages: Prisons, Borders, and Global Crisis*, edited by Jenna Loyd, Matt Mitchelson, and Andrew Burridge, 287–300. Athens, GA: University of Georgia Press.

———. 2014. 'Masculinity and the Mic: Confronting the Uneven Geography of Hip-Hop.' *Gender, Place & Culture* 21 (3): 370–86.

Shaikh, Zeeshan. 2015 (November 21). '"Right to Pee" Activist Mumtaz Shaikh Makes It to BBC's 100 Most Inspirational Women.' *India.com*. www.india.com/news/india/right-to-pee-activist-mumtaz-shaikh-makes-it-to-bbcs-100-most-inspirational-women-725102/ (accessed July 19, 2017).

Shakya, Yogendra B., and Katharine N. Rankin. 2008. 'The Politics of Subversion in Development Practice: An Exploration of Microfinance in Nepal and Vietnam.' *Development Studies* 44 (8): 1214–35.

Sharp, Joanne. 1996. 'Gendering Nationhood: A Feminist Engagement with National Identity.' In *BodySpace: Destabilizing Geographies of Gender and Sexuality*, edited by Nancy Duncan, 91–108. New York: Routledge.

———. 2009. 'Geography and Gender: What Belongs to Feminist Geography? Emotion, Power and Change.' *Progress in Human Geography* 33 (1): 74–80.

Shepherd, Laura J. 2008. *Gender, Violence and Security: Discourse as Practice*. London: Zed Books.

Shillington, Laura J. 2013. 'Right to Food, Right to the City: Household Urban Agriculture, and Socionatural Metabolism in Managua, Nicaragua.' *Geoforum* 44 (January): 103–11. doi:10.1016/j.geoforum.2012.02.006.

Silvey, Rachel. 2004. 'Power, Difference and Mobility: Feminist Advances in Migration Studies.' *Progress in Human Geography* 28 (4): 490–506. doi:10.1191/0309132504ph490oa.

Smith, Barbara. 1980. 'Racism and Women's Studies.' *Frontiers* 5 (1): 48–9.

Smith, Sara. 2009. 'The Domestication of Geopolitics: Buddhist–Muslim Conflict and the Policing of Marriage and the Body in Ladakh, India.' *Geopolitics* 14 (2): 197–218. doi:10.1080/14650040802693382.

———. 2011. '"She Says Herself, 'I Have No Future'": Love, Fate, and Territory in Leh, Jammu and Kashmir State, India.' *Gender, Place & Culture* 18 (4): 455–76.

———. 2012. 'Intimate Geopolitics: Religion, Marriage, and Reproductive Bodies in Leh, Ladakh.' *Annals of the Association of American Geographers* 102 (6): 1511–28. doi:10.1080/00045608.2012.660391.

———. 2013. '"In the Past, We Ate from One Plate": Memory and the Border in Leh, Ladakh.' *Political Geography* 35: 47–59.

Somolu, Orcoluwa. 2007. '"Telling Our Own Stories": African Women Blogging for Social Change.' *Gender and Development* 15 (3): 477–89.

Spain, Daphne. 2014. 'Gender and Urban Space.' *Annual Review of Sociology* 40: 581–98. doi:10.1146/annurev-soc-071913-043446.

Spivak, Gayatri Chakravorty. 1999. *A Critique of Postcolonial Reason*. Cambridge, MA: Harvard University Press.

Staeheli, Lynn A., and Patricia M. Martin. 2000. 'Spaces for Feminism in Geography.' *Annals of the American Academy of Political and Social Science* 571: 135–50.

Staeheli, Lynn A., Eleonore Kofman, and Linda Peake, eds. 2004. *Mapping Women, Making Politics: Feminist Perspectives on Political Geography*. New York: Routledge.

Sultana, Farhana. 2009. 'Fluid Lives: Subjectivities, Gender and Water in Rural Bangladesh.' *Gender, Place & Culture* 16 (4): 427–44.

———. 2011. 'Suffering for Water, Suffering from Water: Emotional Geographies of Resource Access, Control and Conflict.' *Geoforum* 42 (2): 163–72. doi:10.1016/j.geoforum.2010.12.002.

Sun, Wanning. 2008. '"Just Looking": Domestic Workers' Consumption Practices and a Latent Geography of Beijing.' *Gender, Place & Culture* 15 (5): 475–88. doi:10.1080/09663690802300829.

Sundberg, Juanita. 2004. 'Identities in the Making: Conservation, Gender and Race in the Maya Biosphere Reserve, Guatemala.' *Gender, Place & Culture* 11 (1): 43–66.

———. 2014. 'Decolonizing Posthumanist Geographies.' *Cultural Geographies* 21 (1): 33–47. doi:10.1177/1474474013486067.

Swanson, Kate. 2007. 'Revanchist Urbanism Heads South: The Regulation of Indigenous Beggars and Street Vendors in Ecuador.' *Antipode* 39 (4): 708–28. doi:10.1111/j.1467-8330.2007.00548.x.

Swarr, Amanda Lock, and Richa Nagar, eds. 2010. *Critical Transnational Feminist Praxis*. Albany, NY: State University of New York (SUNY) Press.

Sylvester, Christine. 1998. 'Handmaids' Tales of Washington Power: The Abject and the Real Kennedy White House.' *Body & Society* 4 (3): 39–66.

Tan, Qian. 2013. 'Flirtatious Geographies: Clubs as Spaces for the Performance of Affective Heterosexualities.' *Gender, Place & Culture* 20 (6): 718–36. doi:10.1080/0966369X.2012.716403.

Trauger, Amy, Carolyn Sachs, Mary Barbercheck, Kathy Brasier, and Nancy Ellen Kiernan 2010. '"Our Market Is Our Community": Women Farmers and Civic Agriculture in Pennsylvania, USA.' *Agriculture and Human Values* 27 (1): 43–55.

Truelove, Yaffa. 2011. '(Re-)Conceptualizing Water Inequality in Delhi, India through a Feminist Political Ecology Framework.' *Geoforum* 42 (2): 143–52. doi:10.1016/j.geoforum.2011.01.004.

UN, United Nations. 2014. 'Report of the Open Working Group of the General Assembly on Sustainable Development Goals.' United Nations. undocs.org/A/68/970 (accessed February 27, 2017).

UNFPA, United Nations Population Fund. 2007. *State of World Population 2007*. Washington, DC: UNFPA. www.unfpa.org/publications/state-world-population-2007 (accessed February 27, 2017).

Valentine, Gill. 1989. 'The Geography of Women's Fear.' *Area* 21 (4): 385–90.

———. 1993. '(Hetero)sexing Space: Lesbian Perceptions and Experiences of Everyday Spaces.' *Environment and Planning D: Society and Space* 11 (4): 395–413.

———. 2014. *Social Geographies: Space and Society*. New York: Prentice Hall.

van den Berg, Marguerite. 2012. 'Femininity as a City Marketing Strategy: Gender Bending Rotterdam.' *Urban Studies* 49 (1): 153–68. doi:10.1177/0042098010396240.

———. 2013. 'City Children and Genderfied Neighbourhoods: The New Generation as Urban Regeneration Strategy.' *International Journal of Urban and Regional Research* 37 (2): 523–36. doi:10.1111/j.1468-2427.2012.01172.x.

van Geel, Annemarie. 2016. 'Separate or Together? Women-Only Public Spaces and Participation of Saudi Women in the Public Domain in Saudi Arabia.' *Contemporary Islam* 10 (3): 357–78. doi:10.1007/s11562-015-0350-2.

Verma, Ritu. 2014. 'Land Grabs, Power, and Gender in East and Southern Africa: So, What's New?' *Feminist Economics* 20 (1): 52–75. doi:10.1080/13545701.2014.897739.

Waitt, Gordon. 2008. '"Killing Waves": Surfing, Space and Gender.' *Social & Cultural Geography* 9 (1): 75–94. doi:10.1080/14649360701789600.

Waitt, Gordon, Kevin Markwell, and Andrew Gorman-Murray. 2008. 'Challenging Heteronormativity in Tourism Studies: Locating Progress.' *Progress in Human Geography* 32 (6): 781–800. doi:10.1177/0309132508089827.

Walby, Sylvia. 1990. *Theorizing Patriarchy*. Oxford: Basil Blackwell.

Walia, Harsha, and Proma Tagore. 2012. 'Prisoners of Passage: Immigration Detention in Canada.' In *Beyond Walls and Cages: Prisons, Borders, and Global Crisis*, edited by Jenna Loyd, Matt Mitchelson, and Andrew Burridge, 74–90. Athens, GA: University of Georgia Press.

Walker, Margath A. 2005. 'Guada-Narco-Lupe, Maquilarañas and the Discursive Construction of Gender and Difference on the US–Mexico Border in Mexican Media Re-Presentations.' *Gender, Place & Culture* 12 (1): 95–111. doi:10.1080/09663690500083081.

Walsh, Catherine. 2015. 'Life, Nature and Gender Otherwise: Feminist Reflections and Provocations from the Andes.' In *Practising Feminist Political Ecologies: Moving beyond the 'Green Economy,'* edited by Wendy Harcourt and Ingrid Nelson, 101–28. London: Zed Books.

Wangui, Elizabeth Edna. 2014. 'Livelihood Shifts and Gender Performances: Space and the Negotiation for Labor among East Africa's Pastoralists.' *Annals of the Association of American Geographers* 104 (5): 1068–81. doi:10.1080/00045608.2014.924734.

Watts, Michael. 2000. 'Political Ecology.' In *A Companion to Economic Geography*, edited by Eric Sheppard and Trevor Barnes, 257–75. Oxford: Blackwell.

Whitson, Risa. 2011. 'Negotiating Place and Value: Geographies of Waste and Scavenging in Buenos Aires.' *Antipode* 43 (4): 1404–33. doi:10.1111/j.1467-8330.2010.00791.x.

Wilson, Elizabeth. 1991. *The Sphinx in the City: Urban Life, the Control of Disorder, and Women.* Los Angeles, CA: University of California Press.

Wilson, Kalpana. 2015. 'Towards a Radical Re-Appropriation: Gender, Development and Neoliberal Feminism.' *Development and Change* 46 (4): 803–32. doi:10.1111/dech.12176.

Wright, Melissa W. 2006. *Disposable Women and Other Myths of Global Capitalism.* London and New York: Routledge.

———. 2014. 'The *Gender, Place & Culture* Jan Monk Distinguished Annual Lecture: Gentrification, Assassination and Forgetting in Mexico: A Feminist Marxist Tale.' *Gender, Place & Culture* 21 (1): 1–16. doi:10.1080/0966369X.2014.882650.

Wright, Sarah. 2012. 'Emotional Geographies of Development.' *Third World Quarterly* 33 (6): 1113–27. doi:10.1080/01436597.2012.681500.

Yardley, Jim. 2012 (June 14). 'In Mumbai, a Campaign against Restroom Injustice.' *New York Times.* www.nytimes.com/2012/06/15/world/asia/in-mumbai-a-campaign-against-restroom-injustice.html (accessed July 27, 2017).

Young, Iris Marion. 1989. 'Polity and Group Difference: A Critique of the Ideal of Universal Citizenship.' *Ethics* 99 (2): 250–74.

Young, Lorraine. 2003. 'The "Place" of Street Children in Kampala, Uganda: Marginalisation, Resistance, and Acceptance in the Urban Environment.' *Environment and Planning D: Society and Space* 21 (5): 607–27. doi:10.1068/d46j.

Young, Stephen. 2010. 'Gender, Mobility and the Financialisation of Development.' *Geopolitics* 15 (3): 606–27.

Zoomers, Annelies. 2010. 'Globalisation and the Foreignisation of Space: Seven Processes Driving the Current Global Land Grab.' *The Journal of Peasant Studies* 37 (2): 429–77.

Index

Agarwal, B. 161, 163–4, 171
agriculture 110, 127–8, 160–2, 168, 183–4
Alexander, M. 142
alter-geopolitics 145

bars 63
beauty 38–9; advertising 31, 32; nationalism 35; pageants 34, 38–9
Bell, D. and Valentine, G. 19
Bhavnani, K., Foran, J., and Kurian. P. 163–4
biodiversity conservation 158, 167–9
bio-power 41
Black Lives Matter 33, 70, 81
body 11, 21, 26–7, 30–2, 34, 36, 42, 44, 60; beauty 30, 32, 38; bodybuilders 35; as a geographic space 11, 26, 30, 37, 40, 125, 144, 190, 192; modification 32, 36–7, 39, 60, 114
Bondi, L. and Laurie, N. 92
Bondi, L. and Rose, D. 82–3
Brown, M. 11
Browne, K. 33, 65

Browne, K. and Bakshi, L. 103
Butler, J. 21, 27–8, 114

care work 13, 108–9, 124, 191
Chant, S. 93, 124, 166
civil rights 140–2
climate change 13, 165, 168, 184–5
climate justice 183
clothing 34, 39–41, 49, 60–1
code switching 33
coding, the body 32, 35
Colls, R. 60
coloniality of nature 175
combat masculinity (*see militarized masculinity*)
commuting 12
consumption 15, 48, 58, 60, 67; identity 59; race 65; sites of 48, 52, 58, 61, 73
Convention on the Elimination of All Forms of Discrimination Against Women (CEDAW) 136, 138
corporeal geographies 26, 28, 30, 35, 39–41, 42–3, 125, 190; corporeal markers 26, 31, 39, 43–5, 190;

Index

corporeal modification 37; gender 32, 35, 38; race 32
Crenshaw, K. 5
critical geopolitics 132
critical race theory (CRT) 6, 18–19
cultural body 26
cultural geography 49
cultural hegemony 123
cultural spaces 181
culture 21, 37, 48–51, 66, 102, 164, 174
Cuomo, D. 137–8

Development Alternative with Women for a New Era (DAWN) 163
development 14–15, 17, 22, 56, 67, 151, 157, 162, 165–8, 172, 174, 181, 192; and gender 126, 159, 162–4, 178
disability 26, 36–7
Doan, P. 19, 60, 92, 115
domestic masculinities 57–8
domestic violence 3, 9, 136–7, 151
Domosh, M. 7
Domosh, M. and Seager J. 53
dress (*see clothing*)

ecofeminism 13, 139, 160–1
economic strategies 17, 22, 108, 109
emotional political ecologies 178
emotions 60, 181, 192
England, K. and Boyer, K. 115
England, K. and Henry, C. 124
Enloe, C. 147, 149–51
environment 13, 19, 36, 87, 91, 116, 156–8, 161, 164, 167–8, 178, 181, 183–4, 186, 192

exclusion 28, 36, 42, 56, 60, 65, 80, 104, 159, 162, 194
Export Processing Zones (EPZs) 122, 191

Farish, M. 147
fashion 35, 40; veiling 40
fear (*see spaces of fear*)
feminist economic geography 110, 112–13, 121, 128, 191
feminist environmentalism 160–1, 164
feminist geography 4–6, 9, 11, 13, 15, 17, 20, 58, 109, 119, 181, 190, 193
feminist geopolitics 13, 132, 143
feminist political ecology (FPE) 174, 176, 181, 186, 192
feminist political geographers 132, 134, 135, 143, 150, 152
feminist post-structuralism 17
Fenster, T. 79, 80, 84
Fluri, J. 149, 151
Fluri, J. and Dowler, L. 141
forbidden space 84

Gender and Development (GAD) 126, 163–4, 178
gender discrimination 162; segregation 117
gendered divisions of labor 7, 13, 22, 109–12, 181
gender earnings ratio 118
gender equality 162, 166–7
gender roles 15, 27, 30, 34, 49, 72, 86, 101, 109, 115–16, 120, 132, 134, 137
gender-based violence 136–8

gentrification 22, 78, 92, 98, 99; gay enclaves 99–100, 102; and gender 98; and class 100; and race 100
Geographic Information Systems (GIS) 94
geopolitics 66, 125, 132, 138, 143–4, 150, 152, 192
Gibson-Graham, J. K. 17, 112
Gilmore, R. W. 41–2, 113, 142
Global North and Global South 3, 13–14, 18, 22, 79, 93, 114, 124, 127, 165, 172
globalization 15, 22, 109, 113, 119, 123, 125, 127–8, 129, 183, 191; neoliberal 21
Gökarıksel, B. and Secor, A. 40

Hanson, S. 7
Hanson, S. and Pratt, G. 109
HARASSmap 90
heteronormative 20, 54–6, 63, 73, 83, 102, 114–15, 146, 193
heterosexual spaces 7
hip-hop 72–3
Hollaback 70, 90
home 8–9, 11, 12, 26, 29–30, 48, 52, 54, 57, 73–4, 138, 143, 180, 184; feminization of 54, 58; gender 21, 52, 53, 55, 84, 116, 124; political action 96; sexuality 56
homelessness 55
homemaking 52, 57
hooks, b. 5, 54, 57, 190, 193
household 8, 11, 33, 53–4, 108–9, 111, 113, 121, 123, 124–5, 128–9, 162, 164, 170-
housing 92–3, 95, 98, 101

Hovorka, A. 15, 101, 174
Hubbard, P. 63, 93
human rights 31, 80, 156–7
human trafficking 128, 191
human–environment relations 6, 13, 15

identity politics 140, 144
ideology of separate spheres 53
informal sector 22, 110–12
informal workers 95
institutionalized racism 92, 142
International Union for Conservation (IUCN) 168
intersectionality 4–5, 28, 68, 100, 112, 140, 159, 192
intimacy 132, 143
intimate geopolitics 12, 143, 152, 192
intimate violence 151

Jacobs, J. 67
Jacobs, J. and Nash, C. 49

Katz, C. 7, 18, 93, 96, 139
Kern, L. 98–9
Kofman, E. and Peake, L. 132–3

land grabbing 165, 172–3
leisure space 48, 61, 63, 104
LGBTQ 8, 10, 55, 65, 82, 92, 98–9, 102–3; youth 55
liberal feminism 160
liminal space 69, 73
livelihoods (*see economic strategies*)
love 52, 56, 67, 144

Madres de la Plaza de Mayo 139
malls 48, 59, 61, 93, 103

Index

marginalization 16, 36, 43, 109–10, 118, 133, 143, 185
masculinist protection 137
masculinity 28, 32–4, 59, 63–4, 68, 73, 91, 147, 182–3
Massey, D. 7, 110, 139
McDowell, L. 7, 9, 16, 109, 114
McKittrick, K. 5, 142, 178
McKittrick, K. and Peake, L. 31
McLean, J. and Maalsen, S. 69, 72
media 21, 48, 52, 54, 68, 73, 139; gender 72; new media 68, 69–71, 72, 73, 74
microfinance 126
migration 4, 12–13, 22, 41–2, 123–5, 165, 194
military 29–30, 37, 146, 147, 148, 151–2; militarized bodies 150; militarized masculinity 147; militarized spaces 146
Minault, G. 11
mobility 11–13, 70, 85, 123–4; social mobility 67
modernity 41, 67, 149
Mohanty, C. 17–18, 49, 159, 164
Mollett, S. 13, 168, 178
Mollett, S. and Faria, C. 5, 13, 159, 181
motherhood 27, 139; maternal bodies 27
Mountz, A. 43, 125
Muslim women 3, 40, 84, 102

Nash, C. 19, 100
Nash, C. and Gorman-Murray, A. 103
nation 133–4, 144, 147, 174, 192

National Coalition against Domestic Violence (NCADV) 9
National Organization for Women (NOW) 53
nationalism 35, 64, 134
nation-states 133–4
neoliberal urban policy 93–4, 99, 101, 104
neoliberalism 92, 119
new media 68, 69, 70, 72; activism 71; blogging 69
non-governmental organization (NGO) 97, 127

Oberhauser, A. 118
occupational segregation 116
Orientalist 146

Pain, R. 92, 151
Pain, R. and Staeheli, L. 143
Pain, R. and Smith, S. 144
Papanek, H. 11
patriarchy 5, 16, 17, 87, 157, 159, 171, 175, 193
Peake, L. 7
Peake, L. and Rieker, M. 80
performativity 26, 27–8, 38–9
Perry, K. 176, 177–8
Phadke, S. 88
Phadke, S., Khan, S., and Ranade, S. 91
place 6–7, 9, 26, 28, 32, 44, 48, 49, 51, 52, 56, 58, 61, 67, 79, 85, 87, 91, 100, 101, 137, 181, 191; workplace 7, 9, 16, 21, 114, 116, 118, 128, 191
political ecology 6, 13, 158
population control 13, 165–6

post-structural feminism 17, 19
post-colonial feminism 18, 19
post-colonial intersectionality 159
power geometry 139
Pratt, G. 13, 17, 42, 124, 128
Pratt, G. and Rosner, V. 11
Pratt, G. and Yeoh, B. 18
praxis 4, 6, 18, 20, 23, 191
precarious labor 115, 122
pride parades 65
prisons 41–2, 142, 146
privilege 3, 4, 14, 19, 28, 31, 32, 38, 61, 100, 104, 144, 159, 186, 193
productive labor 110, 167
proxy citizenship 145
Puar, J. 146
public space 9, 11, 27, 40, 44–5, 61, 63, 72, 78, 79, 81, 82, 83, 85, 87, 88, 90, 91, 101, 103, 104, 137, 138, 161; as exclusionary 57, 61, 78, 82, 85; and gender 22, 84, 86, 137; privatized 97, 103
public transport 12, 93, 94
Pulido, L. 7, 19, 113, 115, 186

queer geographies 19
queer identity 33, 34, 56, 99, 103
queer space 99

race 4, 5, 9, 18–19, 26, 30, 31, 41, 43, 49, 60–1, 65–6, 68, 82, 84, 86, 91–2, 104, 112–13, 118, 138, 140, 142, 143, 157, 165, 172, 193
racial capitalism 115–16, 186
racial discrimination 157
racial segregation 5, 30, 80, 140, 166

racial vulnerability 91, 110, 142, 159, 171, 175
radical vulnerability 28
rape 2–3, 137, 157; as a crime against humanity 136; as a weapon of war 136, 147–8, 152
relational violence 142
representation 34, 35, 38, 41, 48–9, 68, 70, 72, 133, 144, 164; corporal representation 41, 43
reproductive labor 109–10, 167
reproductive rights 136, 144, 165, 185
resistance 23, 57, 64, 88, 125–6, 128–9, 138, 142, 168, 171, 177, 181, 183, 192
Revolutionary Association of the Women of Afghanistan (RAWA) 70
right to the city 78–9, 80, 82; appropriation 184; and gender 90, 101; participation 80
Rio Plátano Biosphere Reserve (RPBR) 168–71
Rocheleau, D., Thomas-Slayter, B., and Wangari, E. 13, 158
Rose, G. 7

sanitation 93, 96–7, 113, 178–80
sati 50
scale 11–12, 48, 109, 112, 121, 125, 126, 128, 132, 135–6, 138, 140, 145, 150–1, 162, 171, 184, 192
Seager, J. 13, 161
second-wave feminism 15
sex tourism 65–7
sex workers 66, 67, 100, 124, 151

sexual assault 2–3, 9, 91; victim blaming 2–3
sexual relationships and marriage practices 145
sexuality 7, 26, 33, 54–6, 65–6, 72, 88, 112, 138, 140, 143, 146, 152, 190, 192; and space 11, 19
Shabazz, R. 5, 42, 72–3
Shillington, L. 184
Shiva, V. 161
shopping 48, 58–61, 103
slippage 27–8, 34, 35
slums 113, 179–80
SlutWalk 2–3, 90, 193
Smith, S. 12, 143–4
social construction 11; gender and performance 26
social identity 4, 21–2, 109, 122
social reproduction 5, 27, 33, 93, 94–5, 96, 108–9, 110, 112, 121, 132, 141, 171
socialist feminism 162
space 4, 7; private and public spaces 9, 11, 40, 63, 69, 84, 132, 138, 140
spaces of fear 87, 89–90; and masculinity 91; and race 141
spatial access 12, 31
spatial segregation 84; Jim Crow era 141; apartheid 142
spirituality 174
Staeheli, L. and Martin, P. 16
Staeheli, L., Kofman, E., and Peake, L. 138
state 13, 22, 92, 98, 125, 133–4, 139–40, 144, 149, 156–7, 162
structural violence 28, 136, 142
subjectivity 26, 28, 52, 157

Take Back the Night 78, 88, 90
territory 22, 103, 146
time–space compression 139; expansion 139
tourism 31, 58, 61, 65–7, 73, 151, 176
tradition 49, 63, 68, 79, 101–2, 117, 176
transgender 20, 33–4, 60, 115, 145
transnational feminism 19
transnational feminist networks (TFNs) 18
transnational solidarity groups 127, 144
Truelove, Y. 179–80

uneven development 113, 128
United Nations Population Fund (UNFPA) 79
United Nations; International Decade of Women 133; UN Women 120; UN Women's Safe Cities Initiative 78, 90
urban areas 7, 11, 52, 78–9, 85, 99, 102, 124
urban policy 92, 93, 94, 99, 101, 104
urban revitalization 53, 92, 95, 98, 100, 178
urbanization 52, 98, 176

Valentine, G. 87, 115
van den Berg, M. 98, 100
veil/hijab 3, 11, 40, 84
Violence Against Women Act 137

wage gap 113, 116, 118
Waitt, G. 64

Waitt, G., Markwell, K., and Gorman-Murray, A. 66
Walk a Mile in Her Shoes 78, 90
Whitson, R. 43
women and violence 9, 55–6, 70, 78, 85, 87, 90, 96, 136–8, 151, 161, 185
Women in Development (WIN) 162
Women, Culture and Development (WCD) 163
women; African American women 5, 12, 30, 61, 93, 101, 113, 141; Black women 3, 5, 176; Latina women 12, 113; Muslim women 40, 84, 102
women's suffrage 133
Women2Drive Movement 70–1
work 9, 12, 13, 16, 22, 29, 30, 42, 43, 49, 53, 55, 63, 67, 96, 100, 108, 109, 112, 114, 115, 117–18, 121–2, 124, 128–9, 137, 151, 163, 178, 179, 191
World Charter for the Right to the City 80
Wright, M. 43, 100, 122, 181

 Taylor & Francis eBooks

Helping you to choose the right eBooks for your Library

Add Routledge titles to your library's digital collection today. Taylor and Francis ebooks contains over 50,000 titles in the Humanities, Social Sciences, Behavioural Sciences, Built Environment and Law.

Choose from a range of subject packages or create your own!

Benefits for you
- » Free MARC records
- » COUNTER-compliant usage statistics
- » Flexible purchase and pricing options
- » All titles DRM-free.

Benefits for your user
- » Off-site, anytime access via Athens or referring URL
- » Print or copy pages or chapters
- » Full content search
- » Bookmark, highlight and annotate text
- » Access to thousands of pages of quality research at the click of a button.

REQUEST YOUR **FREE** INSTITUTIONAL TRIAL TODAY | **Free Trials Available** We offer free trials to qualifying academic, corporate and government customers.

eCollections – Choose from over 30 subject eCollections, including:

Archaeology	Language Learning
Architecture	Law
Asian Studies	Literature
Business & Management	Media & Communication
Classical Studies	Middle East Studies
Construction	Music
Creative & Media Arts	Philosophy
Criminology & Criminal Justice	Planning
Economics	Politics
Education	Psychology & Mental Health
Energy	Religion
Engineering	Security
English Language & Linguistics	Social Work
Environment & Sustainability	Sociology
Geography	Sport
Health Studies	Theatre & Performance
History	Tourism, Hospitality & Events

For more information, pricing enquiries or to order a free trial, please contact your local sales team:
www.tandfebooks.com/page/sales

 Routledge
Taylor & Francis Group | The home of Routledge books

www.tandfebooks.com